GeoPlanet: Earth and Planetary Sciences

More information about this series at http://www.springer.com/series/8821

Stanisław R. Massel

Internal Gravity Waves in the Shallow Seas

Springer

Stanisław R. Massel
Marine Dynamics
Institute of Oceanology
Sopot
Poland

The GeoPlanet: Earth and Planetary Sciences Book Series is in part a continuation of Monographic Volumes of Publications of the Institute of Geophysics, Polish Academy of Sciences, the journal published since 1962 (http://pub.igf.edu.pl/index.php).

ISSN 2190-5193 ISSN 2190-5207 (electronic)
GeoPlanet: Earth and Planetary Sciences
ISBN 978-3-319-18907-9 ISBN 978-3-319-18908-6 (eBook)
DOI 10.1007/978-3-319-18908-6

Library of Congress Control Number: 2015943450

Springer Cham Heidelberg New York Dordrecht London

Printed on acid-free paper

Springer International Publishing AG Switzerland is part of Springer Science+Business Media (www.springer.com)

Series Editors

Managing Editor

Anna Dziembowska
Institute of Geophysics, Polish Academy of Sciences

Advisory Board

Kon-Kee Liu
Institute of Hydrological
and Oceanic Sciences
National Central University Jhongli
Jhongli, Taiwan

Teresa Madeyska
Research Centre in Warsaw
Institute of Geological Sciences
Warszawa, Poland

Stanisław Massel
Institute of Oceanology
Polish Academy of Sciences
Sopot, Poland

Antonio Meloni
Instituto Nazionale di Geofisica
Rome, Italy

Evangelos Papathanassiou
Hellenic Centre for Marine Research
Anavissos, Greece

Kaja Pietsch
AGH University of Science and
Technology
Kraków, Poland

Dušan Plašienka
Prírodovedecká fakulta, UK
Univerzita Komenského
Bratislava, Slovakia

Barbara Popielawska
Space Research Centre
Polish Academy of Sciences
Warszawa, Poland

Tilman Spohn
Deutsches Zentrum für Luftund
Raumfahrt in der Helmholtz
Gemeinschaft
Institut für Planetenforschung
Berlin, Germany

Krzysztof Stasiewicz
Swedish Institute of Space Physics
Uppsala, Sweden

Ewa Szuszkiewicz
Department of Astronomy
and Astrophysics
University of Szczecin
Szczecin, Poland

Roman Teisseyre
Department of Theoretical Geophysics
Institute of Geophysics
Polish Academy of Sciences
Warszawa, Poland

Jacek Tronczynski
Laboratory of Biogeochemistry
of Organic Contaminants
IFREMER DCN_BE
Nantes, France

Steve Wallis
School of the Built Environment
Heriot-Watt University
Riccarton, Edinburgh
Scotland, UK

Wacław M. Zuberek
Department of Applied Geology
University of Silesia
Sosnowiec, Poland

Piotr Życki
Nicolaus Copernicus Astronomical
Centre
Polish Academy of Sciences
Warszawa, Poland

To my wife Barbara

Preface

The atmosphere and the oceans form a coupled system which constantly exchanges momentum, heat and water through the air-ocean interface. The interface and ocean waters are in endless motions of various scales, from micro-turbulence through to surface waves, tides, global currents and internal waves which are the subject of discussion in this book. In contrast to the surface waves, internal waves propagate through the whole fluid space not only horizontally, but also vertically. In general, ocean fluid is inhomogeneous in space and time due to differences in water temperature and salinity. The surfaces of the interface between warm and cold water or between fresh and salt water in a continuously stratified fluid are unstable. Oscillating interface forms waves which are of different type than those appearing on the free ocean surface. The amplitudes of gravity internal waves within the water column are usually considerably larger; the wavelengths are also much bigger than those of the ordinary surface waves.

Internal waves play a fundamental role for supporting the hydrological and energetic status of oceans. This fact results from the vertical structure of ocean waters. Temperature and salinity of waters differ considerably and cause a substantial variation of the water density. The internal waves generated by wind, pressure fluctuation or flow over the bottom topography, appear to be the principal cause of water mixing. Mean and eddy flows over topography generate internal lee waves, while tidal flows over topography generate internal tides. Vertical mixing enhances meridional heat flux in the ocean, playing an important role in affecting the earth's climate.

While internal waves were known to exist in the ocean in the early 1900s, it is only within the past five decades or so that technology has advanced to a point where large number of observations, also from space, are available. Internal waves move along interfaces as well as in fully three-dimensional space, and they are changing at a relatively fast temporal and small spatial scales, which makes them difficult to observe and include in weather and climate model. Also, the appearance of waves at the boundary surface between two water layers has for long escaped the attention of the observers, because even when the amplitude of the oscillation at the boundary surface is large, the free surface of the upper layer is only slightly disturbed and remains practically at rest.

In contrast to the ocean surface waves, the literature on the internal waves is not so numerous. Due to the high complexity and nonlinear nature of the phenomena, gravity internal waves remain a hydrodynamic process that is still not fully known. Solving the equations describing their evolution poses various mathematical challenges associated with singular boundary value problems and large amplitude dynamics. However, some available oceanographical textbooks and many papers include a discussion of internal gravity waves propagating at interfaces as well as in continuously stratified fluid. Different dynamics mechanisms, such as wave generation, propagation, nonlinear evolution and stability, are also described. The early theoretical results and observations have been summarised in the monographs by Krauss, Roberts and Phillips. In Roberts's monograph, a long list of locations of reported observations of internal waves till the year 1975 is given. In the more recent monograph by Sutherland, the approach taken there is to provide the physics and mathematics describing internal gravity waves in a way that is accessible to students who have been exposed to multivariable calculus and ordinary differential equations. A few books and many papers, mostly theoretical, on the internal gravity waves were also published in Russia, but due to the language barrier, some of these papers are unknown among the Western scientific community. For example, Miropolsky' monograph, originally published in Russia in the year 1982, has been published in English only in the year 2001.

Most of the professional literature dedicated to internal waves is focused on the deep ocean, and only a limited number of papers is dealing with shallow waters. Therefore, there is a need to bring the attention of marine community to specific mechanisms of internal waves propagated in the limiting water depth and over undulating shallow sea bottom. In this book an introduction to the physics and prediction of internal waves in shallow water is presented, taking an advantage of modern theoretical and experimental achievements published in professional literature, along with the author's results. The shallow water case requires a different approach than the deep sea for proper representation of the governing equations describing the internal waves propagation. Therefore, special attention in the book is given to the Korteweg-de Vries equation and its stable solutions in the form of the solitary and cnoidal waves. Another important objective of this book is an examination of various experimental methods for collecting and processing data on the internal waves. In particular, the experimental results obtained by moving sensors and thermistor chains in shallow waters in the Southern Baltic Sea are discussed in detail.

The book is designed for graduate students, Ph.D. students, postdoctoral students and scientists—physicists, engineers and ecologists working on sea environmental problems. The reader is assumed to have a working knowledge of calculus, and a knowledge of hydrodynamic concepts would be helpful. While the fundamental concepts of internal wave physics are given in great detail, some advanced topics are only overviewed. However, many illustrating examples, cross references and subject index are provided to make the book more efficient as a handbook.

The book structure can be summarised as follows: in Chap. 1, useful definitions related to the internal waves are given and the stability of water masses for different profiles of water density is explained. The main purpose of this chapter is to develop general governing equations for various characteristic internal wave parameters.

Chapter 2 deals with the special case of interfacial waves propagating in two-layer seas when the thermocline and pycnocline change very sharply with water depth. The analytical solutions for infinitesimal waves as well as solutions for large amplitude (Stokes' type solution) and long waves (solitary and cnoidal solution) are discussed in detail.

In Chap. 3, a more general case of the continuously stratified ocean is considered for constant water depth. The classical, as well as the variational type solutions of the basic governing equations for short and long internal waves are developed.

Chapter 4 deals with the propagation of internal waves over bottom irregularities and sloping continental shelves. The observation of breaking internal waves in the laboratory channels as well as numerical simulations for propagation of internal waves in a natural environment is described.

Chapter 5 is dedicated to the possible physical mechanisms of generation of the internal waves such as wind action, fluctuation of atmospheric pressure and local disturbances.

In Chap. 6 emphasis is placed on the methodology of the spectral representation of the internal waves in frequency and wavenumber space. In particular, an application of the Fourier Integral or Fourier-Stjeltjes Integral methods for experimental series, stationary in time or uniform is space, is provided. Moreover, using of the Wavelet Transform and Hilbert Transform techniques for non-stationary experimental data is illustrated.

Finally in Chap. 7, practical applications of theoretical methods developed in the previous chapters for the internal waves in the shallow Baltic Sea are described in detail. Various methods of the eigen-solutions of the vertical velocity equation for three-layer model of various density distributions are considered and many examples of the practical calculations are given. The available observations of internal waves with moving sensors and thermistor chains are described and the results of calculations are discussed and interpreted.

The book is supplemented with an extended list of relevant bibliography and subject index. Some figures in the book have been adapted from previously published figures, and this is indicated in the legends. I apologise for the inevitable errors which may occur in this book, despite my efforts to eliminate them. Please bring these errors to my attention.

I would like to express my thanks to the Institute of Oceanology of the Polish Academy of Sciences for support during the preparation of this book. Thanks are also due to Daniel Rak from the Institute of Oceanology for his help in preparation of experimental data from the r/v *Oceania* cruises in the Southern Baltic. I would like to thank my grand-daughter Maria Dembska for her accurate proof-reading and language corrections. I am grateful also to Danuta Pruszczak for her expertise in preparation of the text and figures. Special thanks go to my wife Barbara for her support and encouragement during the writing process.

Sopot
May 2015

Stanisław R. Massel

Contents

Chapter 1
Introducing Internal Waves

1.1 Introduction

Naturally occurring flows involve fluids of different densities, e.g. warm and cold or fresh and saline waters. Under the action of the gravitational forces, the heavy fluid is lowered and the light fluid raises. The ocean is generally stably stratified with the thermocline, below the upper mixed layer which is 100 m deep on average. However, the interface of two media of different densities is usually unstable and under some forcing impulse the wave type disturbances are generated at the interface. A good example of such waves are the ocean surface waves (Massel 2013). However, as the density difference between water and air is large, the interface waves are not too high. On the other hand, within the ocean water column, the vertical density gradients are typically very small and a little energy is required to move water particles vertically. Therefore, the amplitudes of internal disturbances can be very large. For example, Bockel (1962) observed internal waves of 180 m height in the Strait of Gibraltar.

Although coherent oscillations are excited by tides in coastal regions (baroclinic tide), internal wave activity mostly occurs as random fluctuations. The wave energy and statistical properties show remarkable similarities at any time in most part of the ocean, except maybe in the areas near the equator, near strong currents or very shallow waters and over sloping bottom. Various spectral measurements have been summarised in very well known spectral model of Garrett and Munk (1975, 1979).

In a stratified fluid, gravity waves propagate through the fluid vertically as well as horizontally. Such gravity waves moving within a continuously stratified fluid are called **internal waves**, while the term **interfacial waves** is used to describe gravity waves moving at the interface between two water layers of different constant densities. Internal gravity waves transport the momentum and energy, without any transport of matter as long as their dynamics is linear and non-dissipative. Ocean internal waves typically have wavelengths ranging from hundreds of meters to tens of kilometres and periods from several minutes to several hours. Orbital motions of the water particles associated with internal waves have the largest radius at the pycnocline

S.R. Massel, *Internal Gravity Waves in the Shallow Seas*,
GeoPlanet: Earth and Planetary Sciences, DOI 10.1007/978-3-319-18908-6_1

depth, and it decreases downwards and upwards from this depth. The shape of internal waves depends upon the water depth. Near water surface, the commonly appearing waves have a trochoidal form with broad crests and narrow troughs. However, near sea bottom this trend is reversed and the crests, similarly like those of surface waves, are narrower than the troughs (Thorpe 1968; Sutherland 2010).

A role of the internal waves for the ocean's dynamics and ocean's life is underlined in Sect. 1.2. A short review of available observations of the internal waves in the oceans is given in Sect. 1.3. In Sect. 1.4, the basic definitions used for description of internal waves are discussed. For understanding of the physics of internal waves, the stability of ocean water plays a fundamental role. The three states of a fluid element, i.e. stable, unstable or neutral are discussed in detail in Sect. 1.5. In Sect. 1.6, the governing equations for the internal wave characteristics are developed and summarized.

1.2 Internal Ocean Waves and Their Significance

The ocean space is limited by two boundaries, i.e. ocean surface and ocean bottom. Atmospheric forcing at the ocean surface induces surface waves whose energy strongly attenuates with water depth. Therefore, the surface waves play an important role for the energy transfer from the atmosphere, and for the enhancement of heat flux and mixed layer in the upper ocean. As the influence of wind-induced waves is most sensible and hostile on engineering structures, the surface waves are of great interest for the engineering community.

However, at the deeper layers of the ocean, surface waves are not playing a leading role for the fluid motion. Here, the basic sources of energy are tides, currents and internal waves. In particular, ability of internal waves to propagate horizontally as well as vertically provides necessary mechanisms for the transport and disperse of momentum and energy within the total ocean column. Transportation of energy takes place not only in space but also across scales of motions, from low to high frequencies and from low wavenumbers to high wavenumbers.

Although internal waves do not play an important role in the evolution of weather and climate, their influence is non-negligible. They have long been examined as an agent of mixing the ocean to produce the vertical diffusivities needed to maintain the thermoclines in balance with mean vertical advection, contributing to the vertical velocity in the ocean which is required to support the global thermohaline circulation. According to Munk (1966), the vertical velocity of the order of 1.2 cm per day (or 1.4×10^{-7} m s^{-1}) and eddy diffusivity of $K = 1.3$ cm^2 per day are required for that effect. His computations were based on the budget of the cold water (produced in the Antarctica) for the Central Pacific.

In fact, the mechanisms responsible for the generation of the vertical velocity and required diffusivity are still not well known. The four various processes have been considered by Munk, namely boundary mixing, thermodynamic mixing, shear mixing and biological mixing. It is believed that in all these processes, internal waves provide one of the main mechanisms for the mixing in the ocean column, and mixing near boundaries.

Internal waves can also move biological particles and nutrients. They greatly affect the water mixing and transport of gasses and other particulate substances, which provide a necessary biophysical coupling mechanisms for phyto- and zooplankton evolution. Kahru (1983) showed that internal waves mixing and pumping are the important processes that locally raise the nutrient concentration and increase the primary production creating a patchy field of phytoplankton blooms. Lucas et al. (2011) reported measurements from the continental shelf showing that the internal waves transport nitrate shoreward approximately with the same intensity as the rate of phytoplankton nutrient uptake, contributing significantly to the enhanced biological productivity.

The exchange of water between coastal and offshore environments due to internal waves motion is of particular interest for its role in delivering mezoplanktonic larvae to adult populations from offshore larval pools (Botsford et al. 1994). The arrival of cool, deep water associated with internal bores into warm, shallower waters corresponds with drastic increases in phytoplankton and zooplankton concentrations. While both surface waters and those at depth tend to have relatively low primary productivity, thermoclines are often associated with a chlorophyll maximum layer. Those layers attract large aggregations of mobile zooplankton that internal bores push inshore. Many taxa can be almost absent in the warm surface waters but they are plentiful in the internal bores approaching shore (Leichter et al. 1998). In the nonlinear large steep internal waves, the reverse-oscillating cores are observed (Derzho and Grimshaw 1997; Manasseh et al. 1998). These cores can also transport parcels of water shoreward with suspending sediment as well as plankton and nutrients found along the benthos in deeper water.

Internal waves induce oscillation of thermoclines and therefore have the potential to transfer the phytoplankton-rich waters downward from pelagic to benthic system. In particular, periodic depression of the thermocline and associated downwelling play an important role in the vertical transport of planktonic larvae.

1.3 Experimental Insights into Internal Waves

John Scott Russell, working in the 19th century, was probably the first to observe internal waves (Russell 1844). He reported the formation of a single unchanging mound in the shallow waters of the narrow channel, generated when a towed large barge was suddenly stopped. As Russell noted, a rounded, well-defined heap of water continued to move without change of form and speed. The observation of such solitary waves and the subsequent theoretical description by Korteweg and de Vries (1895) provided some understanding of this phenomenon at the beginning of 20th century.

However, it is only within the last decades or so that technology has advanced to a point where large numbers of observations become available. Most internal wave measurement methods are based on recording the profiles of temperature and salinity in the water column. The movement of points of equal temperature or the movement of points of equal salinity are manifestations of the passage of internal waves.

Since the launching of the *Seasat* in 1978 and *European Remote Sensing Satellite ERS-1* in 1991 with a synthetic aperture radar (SAR) on board, a large number of data on the surface manifestation of the presence of internal waves has become available. Historically, the Andaman Sea was probably the region where the surface effects of internal waves passage were first observed (Osborne and Burch 1980). Based on satellite photography, the crest lengths of the observed internal solitons may be as long as 150 km and the separation distance between solitons in a packet as long as 15 km. Packets occurred every 12 h 26 min, indicating a tidal origin for the internal solitons. The source regions for the solitons generation are shallow-water areas near northern Sumatra or the southernmost of the Nicobar Islands, where the average spring tide range is $\sim$1.4 m.

The large amplitudes of internal waves (up to 100 m or more) have been observed in Sulu Sea (Apel et al. 1985), at the Mascarene Ridge (Sabinin 1992; Konyaev et al. 1995), near Gibraltar Strait (Bockel 1962; Bryden et al. 1994) and in the central Bay of Biscay (New and Pingree 1990).

The internal waves of tidal frequency (internal tides) are generated by the interaction of barotropic tidal currents with sloping topography in stratified fluid at the critical points where slope of the internal wave characteristics is the same as the bottom slope (see Chap. 4 for more details). From these critical points, known as generation sites, the beams of energy propagate in both shoreward and seaward direction. During shoreward propagation the wave packet is continuously modified and the wave amplitude grows. Comprehensive observations of internal tides were made by Holloway (1984, 1994, 1996) on the semi-diurnal internal tides in the region of the shelf-break at the Australian North West Shelf. On the shelf, tidal amplitudes range up to 10 m and the magnitude of the dissipation of tidal energy is very high. It was shown that the internal tides have a three-dimensional structure and the amplitude at the M_2 tidal frequency dominates over the S_2 amplitude (Burrage et al. 1996). Predominantly the waves of depression are observed during the summer, while in winter they are weaker and are predominantly waves of positive elevation (Van Gastel et al. 2009).

Van Gastel et al. accurately measured the phase speed and direction of the propagation of the internal wave packet using the current meter array and L-shaped benthic array containing six cross-shelf and long-shelf moorings along two orthogonal lines. All instruments collected temperature and current magnitude and direction measurements enabling both the phase speed and direction of the internal waves. The observed phase speeds of the waves, calculated from thermistor chain and current meter data are shown to be in a close agreement with the linear theory, providing that slowly varying depth is taken into account.

The Indonesian Throughflow which joins the waters of the Pacific and Indian Oceans has been a subject of very large investigations from the 1990s. The system of sub-surface pressure gauges and ADCPs provided the data on the vertically integrated transport. The amplitude of the total surface transport variation through the five major passages is ranging from 10–15 Sv (Chong et al. 2000). Sea levels and mooring data, processed using the Hilbert-Huang Transform (Huang and Shen 2005), display inter-seasonal variability which corresponds to the remotely forced Kelvin waves from the Indian Ocean through Lombok Strait (Dwi Susanto et al. 2000). In particular, the Lombok Strait separating the Indonesian islands of Bali and Lombok is one of the straits in the Indonesian archipelago where high-amplitude internal waves are observed (Dwi Susanto et al. 2005). The analysis of two consecutive satellite SAR images showed that the observed internal waves are a result of the interaction between semi-diurnal tidal flows over the sill, south of the Lombok Strait. Clearly a two-layer system of water column was observed. For solitary internal waves, the horizontal velocities in the lower layer were opposite in direction to those in the upper layer. The echogram showed very large internal waves with amplitudes exceeding 100 m and with propagation speed of 1.96 m s^{-1}.

In recent years, the northern South China Sea (SCS) became a popular *hot spot* for observations of the large internal solitary waves (Lien et al. 2014). They are formed by conversion from internal tides generated in the Luzon Strait and propagate on the upper continental slope in the SCS close to Dongsha Island. Experiments in May 2011 showed that the vertical displacement of internal tides has a nonlinear non-sinusoidal form and amplitude of the displacement is of the order of 50 m. The propagation speed of the waves was about 2 m s^{-1}. Vertical profiles of down welling and upwelling have maxima occurring at around 150 m water depth. Some of the observed internal waves reach the breaking limit $u_{max} > C_0$, indicating convective instability (u_{max} is the maximal horizontal velocity and C_0 is the wave phase speed). Vertical overturns of ~100 m occur within the core of the wave.

Since the early 1980s, the Joint Canada–US Ocean Wave Investigation Project and SAR Internal Wave Signature Experiment provided results widely used in the interpretation and application of ocean SAR images taken by the following missions, such as ERS-1, namely SAR images of internal waves evolution in the East and South China Seas and RADASAT images of internal waves in the Middle Atlantic Bight. Zheng et al. (2001) examined the variation of thermocline depth within an ocean area between 73°–74°W and 38°–39°N. Along the track, water depths increase from 20 to 1800 m. It was observed that the thermocline depth of the upper ocean is shoaling and the density difference between the upper and lower layers varies with bottom shoaling. In particular, the density difference decreases from 3.2×10^{-3} to 1.6×10^{-3} g cm^{-3} within a distance of 60 km. Thermocline shoaling and deepening induces changes in the phase speeds of internal waves as described by linear wave theories.

The variation of the wave speed may reflect the fact that waves are propagating in an inhomogeneous medium. Such evolution of internal waves was observed in the Gulf of Aden from *Discovery* space shuttle during mission STS-26 (Zheng et al. 2001). From the photographs, groups of long curved lines, representing the signatures of ocean internal waves packets can be seen. Some of them indicate that a surprisingly sharp recess of internal solitary wave is a signature of spatial phase delay caused by thermocline shoaling over the seamount.

Serebryanyj (1990) showed a series of examples of internal waves observed in the shelf zones of Kaspiyan, Black and Japan Seas. These observations illustrate a large influence of the nonlinearities on the waves, especially in terms of their asymmetry.

The White Sea is the smallest sea of the Arctic Ocean with a complex bottom topography and coastline. The tidal current velocities are very high, up to 1.8 m s^{-1}. Such conditions are suitable for the generation of the nonlinear internal waves. Kozlov et al. (2014) examined the satellite Envisat Advanced SAR images of the internal waves activity in the particular regions of the White Sea. These internal waves were characterised by the crest lengths as long as 200 km with the wavelengths of leading waves up to 2.5 km. The numerical model, based on the extended Korteweg-de Vries equation (see Chaps. 3 and 4), confirmed a strong influence of M_2 tidal cycle.

Fluid motion induced by various causes (see Chap. 5) disturbs initial equilibrium, generating isopycnal displacements. Generally speaking, the patterns of these oscillations are non-stationary in time and non-uniform in space. Moreover, in shallow waters, irregular bathymetry enhances the nonlinear mechanisms which cause the random disturbances of the thermoclines and pycnoclines. Therefore, it is sometimes very difficult to describe the internal waves properties by the deterministic models discussed in Chaps. 1–4. The only way is to apply the spectral methods which are given in Chap. 6.

The Baltic Sea was omitted in our discussion. However, the separate Chap. 7 is dedicated to this shallow water basin with the special features in terms of the internal waves dynamics.

1.4 Useful Definitions

For the purpose of this book, we define the **short-period internal waves** as those waves whose periods are significantly shorter than 12 h. Usually, the periods of short internal waves are ranging between 5 and 20 min, being as short as 2.5 min or long as 5 h. For example, temperature measurements made in Massachusetts Bay (West Atlantic) indicate that the thermocline can heave up and down with periods of 6 to 8 min with vertical displacements of 10 m, occurring at 17.2 m below the surface (Halpern 1971). In the Andaman Sea, short internal waves with periods of about 20 min and amplitudes over 40 m were observed.

Internal waves with periods anywhere around 12 h are called **semi-diurnal tides**, and internal waves with periods around 24 h are called **diurnal internal tides**. The amplitudes of internal tides are usually 2–10 m, while their length is about 3×10^4 m. A comprehensive review of internal tide observations was published by Huthnance (1989). In particular, the existence of internal waves with a semi-diurnal period was observed during temperature/salinity measurements on the Australian North-West Shelf. Vertical displacements of isopycnal surfaces measured about 30 m, equal to nearly half the water depth (Holloway 1983). Internal waves with the longest periods are **inertial waves**, which for our purpose are defined as waves whose periods are very close to the inertial period for a particular latitude.

In contrast to the internal wave amplitude, it is very difficult to determine experimentally the dispersion relation and wavelength. The phase speed of the short-period internal waves varies between 0.1 and 1 m s^{-1}. Steepness of waves is very small, equal to about 0.04. They are not necessary of sinusoidal form. Sometimes they are flattened on the crests, when the thermocline is shallow and peaked, when the thermocline is deep.

Seawater is not pure water. It contains salts, dissolved organic substances, and mineral and organic suspended matter. The principal seawater state parameter is **density**, or mass per unit volume. The density of water controls many fundamental processes in the ocean, e.g. the hydrostatic equilibrium or motion of water particles and propagation of internal waves, turbulence and mixing in water column, heat transfer, concentration of plankton and sediment transport, locomotion of marine species and many others. Even small alterations in the density of seawater result in great changes in water flow and its thermal and chemical status.

Density is usually symbolised by ρ and in the SI system it is expressed in kilograms per cubic meter (kg m^{-3}). For example, the density of pure water is approximately 1000 kg m^{-3} being 770–890 times that of air at sea level. The water density is frequently denoted in σ_t units as a form of shorthand. The quantity σ_t known as the potential density is related to density ρ by the expression:

$$\sigma_t = \rho - 1000 \tag{1.1}$$

Thus, the potential density $\sigma_t = 25$ kg m^{-3} corresponds to the density ρ equal to 1025 kg m^{-3}.

Unlike measuring salinity, temperature and pressure, there are no practical methods of measuring density of seawater *in situ*. Density of seawater is usually determined through its dependence on pressure, p, temperature, T, and salinity, S. Pressure, p, has an insignificant effect on the density of water for most applications, unless one is dealing with water at great depths within the ocean.

Density is much more dependent on temperature, generally decreasing as the temperature increases. However, density of freshwater is not a monotonic function of the temperature, and water density reaches a maximum at temperature of 3.98 °C under normal atmospheric pressure. Normal atmospheric pressure at sea level, called the *normal atmosphere*, is defined to be 1.01325×10^5 Pa at 15 °C, in which 1 Pa is equal to 1 N m^{-2}. This anomalous dependence of the density of freshwater on temperature is a result of the specific molecular structure of water.

The general dependence of the sea water density $\rho(S, T, p)$ on salinity, S, temperature, T, and pressure, p, is called the **equation of state** for sea water. There have been many attempts to establish the relationship $\rho(S, T, p)$ in the past and the modern algorithm of determination of density for given salinity, S, temperature, T, and pressure, p, was developed by an international group of experts on standards (UNESCO 1981) see Appendix A.

Most internal waves measurement methods are based on recording of the profiles of temperature and salinity in the water column. The movement of points of equal temperature, **isotherms**, or the movement of points of equal salinity, **isohalines**, as well as the movement of points of equal density, **isopycnals** are manifestations of the passage of internal waves. The zone where the vertical density gradient is the greatest is known as a **pycnocline**. For completeness we define here the **thermocline** as a zone where density is determined mostly by the rapid temperature variations, and **halocline** where the density variations are controlled mostly by the salinity. Figure 1.1 illustrates typical vertical profiles of the temperature, salinity and density in the Southern Baltic waters. The thermocline and halocline are about 60 m below the sea surface, with the isopycnals changing between water depths 50 m and 72 m. In the same figure, a vertical profile of the Väisälä-Brunt frequency, $N(z)$, is also given. We will discuss the frequency N later in this chapter in a detail.

When the density of fluid is a function of pressure only (i.e. $\rho = \rho(p)$), the isobaric and isopycnal surfaces are parallel to each other. This situation is known as a **barotropic** water mass. If density is a function of other parameters (i.e. $\rho = \rho(S, T, p)$), the isobaric and isopycnal surfaces may be inclined to each other. This is known as a **baroclinic** water mass. The barotropic case is most common in deep water, while the baroclinic case is typical for the upper layer of the ocean.

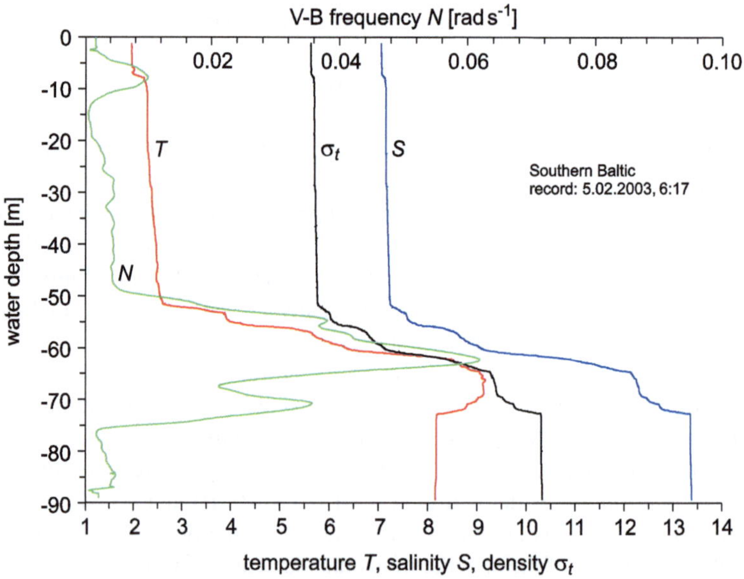

Fig. 1.1 Vertical profiles of water temperature T, salinity S, density σ_t and Väisälä-Brunt frequency N in Southern Baltic

1.5 Stability of Water Masses

Understanding the stability of ocean water is the basic requirement needed to understand the physics of internal waves. In general, a fluid element can be in one of the three states: stable, unstable, or neutral stability. Let us, following Massel (1999), apply these three stability states to a small element of water in the ocean with some specific vertical density profiles (see Fig. 1.2). In the stable case **a**, it is assumed that water density increases with water depth. Initially at position 1, the forces acting on a small element of water, i.e. weight of a given volume of water and buoyancy force are in balance; thus:

$$\rho(z_1)Vg - W = 0 \tag{1.2}$$

in which $\rho(z)$ is the water density at the level $z = z_1$, V is the volume of the small element of the water under consideration, and W is the weight of this element of water.

Suppose now that due to some external force, the water element is moved to position 2, at the level z_2. The weight of the water element remains the same, however, the buoyancy force changes, because of the change of the density of surrounding water. Thus we have:

$$\rho(z_2)Vg - W = \rho(z_2)Vg - \rho(z_1)Vg = [\rho(z_2) - \rho(z_1)]\, Vg < 0 \tag{1.3}$$

Because $\rho(z_2) < \rho(z_1)$, the resulting force is directed downwards and the water element tends to sink back to its original level. A similar consideration for the water element moved to position 3 (level z_3) yields the conclusion that an excess buoyancy force moves the element back to its original position. Thus, case **a** is related to the

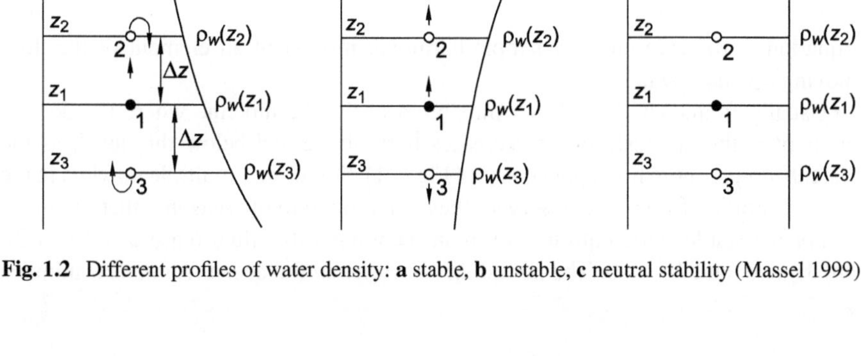

Fig. 1.2 Different profiles of water density: **a** stable, **b** unstable, **c** neutral stability (Massel 1999)

situation defined as stable. A water element approaching its initial position oscillates around that position, until kinetic energy of the element dissipates.

In the unstable case **b**, the water density decreases with depth. Hence, the balance of forces for the water element moved to level z_2 is as follows:

$$\rho(z_2)Vg - W = [\rho(z_2) - \rho(z_1)]\, Vg > 0 \tag{1.4}$$

Thus, the water element will be pushed up, away from its initial position. On the other hand, when the water element is moved from its initial position to position 3 at the level z_3, the buoyancy is not sufficient to balance the element's weight. As a result, the element will sink, moving away from its initial position.

In the neutral case **c**, the water density remains constant; thus the balance of forces is satisfied at any position. Such situation is defined as neutral stability.

Consider now a stable profile shown in Fig. 1.2, and assume that the Earth's rotation is ignored. From Eq. (1.3), it follows that when a small amount of water is displaced from its initial position, the resulting force tends to move it back to the initial position. The movement of the water element is restrained by an inertia force, and the resulting balance of force can be expressed as follows:

$$\rho\frac{d^2\zeta}{dt^2} + \left(-\frac{g}{\rho}\frac{d\rho}{dz}\right)\zeta = 0 \tag{1.5}$$

in which ζ is the vertical water displacement or:

$$\frac{d^2\zeta}{dt^2} + N^2(z)\zeta = 0 \tag{1.6}$$

in which the quantity $N(z)$ is known as the **Väisälä-Brunt frequency** or stability frequency:

$$N(z) = \left(-\frac{g}{\rho}\frac{d\rho}{dz}\right)^{1/2} \tag{1.7}$$

The elementary solution of Eq. (1.6) takes the form (Massel 1999):

$$\zeta(t) = \cos(N\,t) \tag{1.8}$$

Equation (1.8) represents the simple harmonic motion of an element of the fluid moving up and down.

Usually frequency $N(z)$ has a maximum in the thermocline where the density gradient is the greatest, and it decreases both above and below this level, as the water becomes more homogeneous (see Fig. 1.1). Sometimes multiple maxima in the vertical profile of $N(z)$ are observed; they correspond to the seasonal thermocline.

For the stable static equilibrium in an incompressible fluid (case **a** in Fig. 1.2), $\partial\rho/\partial z < 0$, and $N^2 > 0$. The period $2\pi/N$ varies in the ocean from a few minutes in the thermocline, to several hours in the deep ocean. In some circumstances, such as in lakes and in the deep ocean where $\partial p/\partial z$ is very small, the extra term $-g^2/C_s^2$ (where C_s is the sound speed in the ocean) should be included for the Väisälä-Brunt frequency N^2.

In case **b** (Fig. 1.2), when $\partial \rho/\partial z > 0$, and $N^2 < 0$, Eq. (1.6) has no solution of the wave type. In fact, the only solution is:

$$\zeta(t) = \cosh\left[\left(\frac{g}{\rho}\frac{d\rho}{dz}\right)^{1/2} t\right] \tag{1.9}$$

and as t increases, the displacement goes exponentially to infinity. When $N = 0$, the water mass is uniform in terms of density.

As will be shown in the following chapters, the frequency N plays a fundamental role in determining internal wave patterns.

1.6 Governing Equations

For convenience, a short presentation of the basic equations for internal waves is given. We assume a Cartesian coordinate system x, y, z (z axis positive upward) with unit vectors **i**, **j**, **k**. The instantaneous velocity components are u^*, v^*, w^* in x, y, and z, respectively. Therefore, the total instantaneous velocity vector may be written as follows:

$$\mathbf{u}_t^* = u^*\mathbf{i} + v^*\mathbf{j} + w^*\mathbf{k} \tag{1.10}$$

Let us discuss the continuity and momentum equations (Roberts 1975; Massel 1999) i.e.:

$$\frac{D\rho^*}{Dt} + \rho^*(\nabla \cdot \mathbf{u}_t^*) = 0 \tag{1.11}$$

in which:

$$\frac{D}{Dt} = \frac{\partial}{\partial t} + u^*\frac{\partial}{\partial x} + v^*\frac{\partial}{\partial y} + w^*\frac{\partial}{\partial z} \tag{1.12}$$

and

$$\rho^*\frac{D\mathbf{u}_t^*}{Dt} + 2\rho^*(\Omega_v \times \mathbf{u}_t^*) + \nabla p^* + g\rho^*\mathbf{k} = 0 \tag{1.13}$$

where $\Omega_v = \Omega_x\mathbf{i} + \Omega_y\mathbf{j} + \Omega_z\mathbf{k}$ is the angular velocity of the Earth rotation. In Eqs. (1.11)–(1.13) incompressible fluid and absence of outside forces is assumed.

We assume now that sea water is incompressible and diffusion is neglected. Thus we have:

$$\frac{D\rho^*}{Dt} = 0 \tag{1.14}$$

As a consequence of Eq. (1.14), the continuity equation (1.11) becomes:

$$\nabla \cdot \mathbf{u}_t^* = 0 \tag{1.15}$$

For further convenience we represent the basic quantities as sum of a mean, time-independent background component and perturbation component, resulting from wave motion, i.e.:

$$\rho^* = \overline{\rho}(z) + \rho(x, y, z, t) \tag{1.16}$$

$$p^* = \overline{p}(x, y, z) + p(x, y, z, t) \tag{1.17}$$

$$u^* = U(x, y, z) + u(x, y, z, t) \tag{1.18}$$

$$v^* = V(x, y, z) + v(x, y, z, t) \tag{1.19}$$

$$w^* = W(x, y, z) + w(x, y, z, t) \tag{1.20}$$

Using the representations above, let us rewrite the important density equation (1.14) as follows:

$$\frac{\partial}{\partial t}(\overline{\rho} + \rho) + u\frac{\partial}{\partial x}(\overline{\rho} + \rho) + v\frac{\partial}{\partial y}(\overline{\rho} + \rho) + w\frac{\partial}{\partial z}(\overline{\rho} + \rho) = 0 \tag{1.21}$$

For stationary and horizontally uniform background components we have $\partial\overline{\rho}/\partial t = \partial\overline{\rho}/\partial x = \partial\overline{\rho}/\partial y = 0$. The nonlinear terms in the equation above are negligible for small amplitude motion. However, the linear term $w\partial\overline{\rho}/\partial z$ must be retained. Thus, Eq. (1.21) can be simplified as follows:

$$\frac{\partial \rho}{\partial t} + w\frac{\partial \overline{\rho}}{\partial z} = 0 \tag{1.22}$$

This equation demonstrates the important fact that the density perturbation at a given point is generated only by a vertical advection of the background density distribution. As $w = \partial\zeta/\partial t$, we have:

$$\frac{\partial}{\partial t}\left(\rho(x, z, t) + \zeta(x, z, t)\frac{\partial \overline{\rho}(z)}{\partial z}\right) = 0 \tag{1.23}$$

Therefore, the water density $\rho(x, z, t)$ becomes:

$$\rho(x, z, t) = -\zeta(x, z, t)\frac{\partial\overline{\rho}(z)}{\partial z} \tag{1.24}$$

Let us present the total velocity $\mathbf{u}_t^*$ in the form:

$$\mathbf{u}_t^* = u^*\mathbf{i} + v^*\mathbf{j} + w^*\mathbf{k} = (U + u)\mathbf{i} + (V + v)\mathbf{j} + (W + w)\mathbf{k} = \mathbf{U} + \mathbf{u} \tag{1.25}$$

where:

$$\mathbf{U} = U\,\mathbf{i} + V\,\mathbf{j} + W\,\mathbf{k} \tag{1.26}$$

and

$$\mathbf{u} = u\,\mathbf{i} + v\,\mathbf{j} + w\,\mathbf{k} \tag{1.27}$$

Thus, the momentum Eq. (1.13) becomes:

$$(\overline{\rho} + \rho)\frac{D\mathbf{u}_t^*}{Dt} + 2(\overline{\rho} + \rho)(\Omega \times \mathbf{u}_t^*) + \nabla(\overline{p} + p) + g(\overline{\rho} + \rho)\mathbf{k} = 0, \tag{1.28}$$

In the absence of motion and outside forces, the vertical component of Eq. (1.13) is:

$$\frac{\partial\overline{p}}{\partial z} + g\overline{\rho} = 0 \tag{1.29}$$

which is the equation of hydrostatics.

In this book the assumption of the Boussinesq fluid is frequently used. The Boussinesq approximation is invoked if the density of the stratified fluid in which the internal waves propagate varies little over the fluid depth. In fact, in the ocean, the density varies by only a few per cent from surface to bottom. In the Boussinesq approximation, the density fluctuations are negligible except for the case when they affect buoyancy forces (see term $g(\overline{\rho} + \rho)\mathbf{k}$ in Eq. 1.28). Using this assumption we have from Eq. (1.13) (Roberts 1975):

$$\frac{D(\mathbf{U} + \mathbf{u})}{Dt} + 2\Omega_v \times (\mathbf{U} + \mathbf{u}) + \frac{\nabla p}{\overline{\rho}} + \frac{g\rho\mathbf{k}}{\overline{\rho}} = 0 \tag{1.30}$$

Let us now neglect the nonlinear terms assuming that $\overline{u}^*$ and $\nabla\overline{u}^*$ are small, so their product is smaller jet. Also, we take $\mathbf{U} = 0$, $\Omega_x = \Omega_y = 0$, $2\Omega_E \sin\varphi = f$, where $\Omega_E = 7.29 \times 10^{-5}$ rad s^{-1} is angular frequency of Earth rotation and φ is a latitude of a given point. The frequency f is known as the inertial frequency.

Therefore, Eqs. (1.30) and (1.22) give:

$$\frac{\partial u}{\partial t} - fv + \frac{1}{\overline{\rho}}\frac{\partial p}{\partial x} = 0 \tag{1.31}$$

$$\frac{\partial v}{\partial t} + fu + \frac{1}{\overline{\rho}}\frac{\partial p}{\partial y} = 0 \tag{1.32}$$

$$\frac{\partial w}{\partial t} + \frac{1}{\overline{\rho}}\frac{\partial p}{\partial z} + g\frac{\rho}{\overline{\rho}} = 0 \tag{1.33}$$

$$\frac{\partial \rho}{\partial t} + w\frac{\partial \overline{\rho}}{\partial z} = 0 \tag{1.34}$$

$$\frac{\partial u}{\partial x} + \frac{\partial v}{\partial y} + \frac{\partial w}{\partial z} = 0 \tag{1.35}$$

Above important equations are used in the following chapters of the book.

Roberts (1975) presented Eq. (1.30) in terms of the vertical velocity w as:

$$\frac{\partial^2}{\partial t^2}\left(\nabla^2 w\right) + f^2\frac{\partial^2 w}{\partial z^2} + N^2(z)\nabla_h^2 w - \frac{N^2(z)}{g}\left(\frac{\partial^3 w}{\partial z \partial t^2} + f^2\frac{\partial w}{\partial z}\right) = 0 \tag{1.36}$$

However, for further derivations we apply representation given by Phillips (1977):

$$\frac{\partial^2}{\partial t^2}\left(\nabla^2 w\right) + f^2\frac{\partial^2 w}{\partial z^2} + N(z)^2\nabla_h^2 w = 0 \tag{1.37}$$

For some applications it is useful to represent the Eq. (1.37) in terms of the stream function $\psi(x, z, t)$ or velocity potential function $\Phi(x, z, t)$ defined as:

$$w = -\frac{\partial \psi}{\partial x} \quad \text{and} \quad u = \frac{\partial \psi}{\partial z} \tag{1.38}$$

or

$$w = \frac{\partial \Phi}{\partial z} \quad \text{and} \quad u = \frac{\partial \Phi}{\partial x} \tag{1.39}$$

Therefore Eq. (1.37) becomes:

$$\frac{\partial^2}{\partial t^2}\left(\nabla^2 \psi\right) + f^2\frac{\partial^2 \psi}{\partial z^2} + N^2(z)\frac{\partial^2 \psi}{\partial x^2} = 0 \tag{1.40}$$

or

$$\frac{\partial^2}{\partial t^2}\left(\nabla^2 \Phi\right) + f^2 \frac{\partial^2 \Phi}{\partial z^2} + N^2(z) \frac{\partial^2 \Phi}{\partial x^2} = 0 \tag{1.41}$$

If the vertical velocity w is periodic in time with frequency ω, we have:

$$w(x, y, z, t) = W(z)\, F(x, y)\, \exp(-i\omega t) \tag{1.42}$$

After substituting above representation into Eq. (1.37) and separating the variables we obtain the following set of equations:

$$\frac{d^2 W}{dz^2} - \frac{N^2(z)}{g} \frac{dW(z)}{dz} + (N^2(z) - \omega^2)\nu W = 0 \tag{1.43}$$

and

$$\nabla_h^2 F + (\omega^2 - f^2)\nu F = 0 \tag{1.44}$$

where:

$$\nu = \frac{k_h^2}{\omega^2 - f^2} \tag{1.45}$$

in which:

$$k_h^2 = k_x^2 + k_y^2 \tag{1.46}$$

Following Krauss (1966) we assume that $\left|\frac{d^2 W}{dz^2}\right| \gg \left|\frac{N^2}{g} \frac{dW}{dz}\right|$. Thus Eq. (1.43) simplifies as follows:

$$\frac{d^2 W}{dz^2} + \frac{N^2(z) - \omega^2}{\omega^2 - f^2} k_h^2\, W = 0 \tag{1.47}$$

Defining:

$$k_z^2 = \frac{N^2(z) - \omega^2}{\omega^2 - f^2} k_h^2 \tag{1.48}$$

Equation (1.47) becomes:

$$\frac{d^2 W}{dz^2} + k_z^2 W = 0 \tag{1.49}$$

If $k_z^2 < 0$, the solution of Eq. (1.49) is exponential in z, and resulting motion is of the surface-trapped wave. However, for a positive k_z^2, solution has a trigonometric form for the internal waves.

If the period of waves is much shorter than the inertial period, Eq. (1.47) can be further simplified as:

$$\frac{d^2 W}{dz^2} + \frac{N^2(z) - \omega^2}{\omega^2} k_h^2 W = 0 \tag{1.50}$$

For the prescribed Väisälä-Brunt frequency, $N(z)$, the frequency of internal waves, ω, cannot be arbitrary. At a given depth, this frequency is bounded above by the Väisälä-Brunt frequency, $N(z)$, and below by the inertial frequency f, i.e.:

$$f^2 < \omega^2 < N^2(z) \tag{1.51}$$

The inertial frequency varies from about 1.45×10^{-4} rad s^{-1} at the poles ($T_f \sim 12$ h) to 0 at the Equator (T_f infinite). The inertial frequency is the minimum frequency that the free internal wave motion can possess. Assuming that $N(z)$ may be as large as 0.01 rad s^{-1} ($T_N \approx 10$ min), we obtain the band of possible internal wave frequencies as shown in Fig. 1.3.

At a given depth z, the Väisälä-Brunt frequency, $N(z)$, determines the maximal frequency for the waves which occur at that depth. Since $N(z)$ is greatest in the thermocline, where the variation of density with depth is greatest, internal waves of the largest frequency occur in the thermocline.

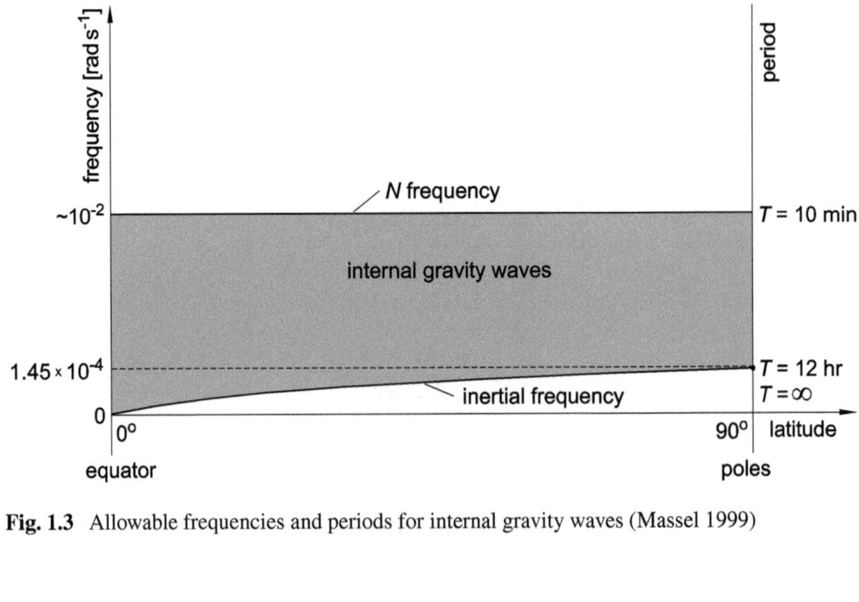

Fig. 1.3 Allowable frequencies and periods for internal gravity waves (Massel 1999)

Let us discuss now the dispersion relation of the internal waves in some detail, neglecting the Earth's rotation for simplicity. From Eq. (1.48) we have:

$$\omega = \frac{k_h N(z)}{\sqrt{k_x^2 + k_y^2 + k_z^2}} = \frac{k_h N(z)}{K} \tag{1.52}$$

or

$$\omega = \pm N(z) \cos\theta \tag{1.53}$$

in which $K^2 = k_x^2 + k_y^2 + k_z^2$ and θ is the angle between the wave vector $\mathbf{K}$ and the horizontal plane. It should be noted that the frequency of the internal wave depends only on the direction of the wave vector and not on its magnitude, which is in the clear contrast to the surface gravity waves for which frequency depends only on the wavenumber magnitude.

The frequency of internal waves is therefore independent of the wavelength. Two signs are allowed for internal waves to propagate upward or downward along the wavenumber direction. If frequency is fixed, all waves, regardless of wavelength, propagate at fixed angles from the horizontal plane. The lower frequency ($\omega \ll N$), the steeper the direction, when waves propagate almost vertical ($\theta \approx \pi/2$). The higher frequency ($\omega \approx N$), propagation angle is close to horizontal.

If the inertial frequency f is taken into account, from Eq. (1.48) we obtain dispersion relation in the form (motion in vertical plane (x, z) was assumed):

$$\omega^2 = N^2(z) \frac{k_x^2}{k_x^2 + k_z^2} + f^2 \frac{k_z^2}{k_x^2 + k_z^2} \tag{1.54}$$

or

$$\omega^2 = N^2(z) \cos^2\theta + f^2 \sin^2\theta \tag{1.55}$$

Therefore, an angle of inclination of wave vector lines in the vertical plane (x, z) becomes:

$$\theta = \arccos\left(\frac{\omega^2 - f^2}{N^2(z) - f^2}\right)^{1/2} \tag{1.56}$$

Let the velocity field takes the form:

$$(u, v, w) = (U, V, W) \exp[i(k_x x + k_y y + k_z z)] \tag{1.57}$$

then the continuity equation imposes the condition:

$$k_x U + k_y V + k_z W = 0 \tag{1.58}$$

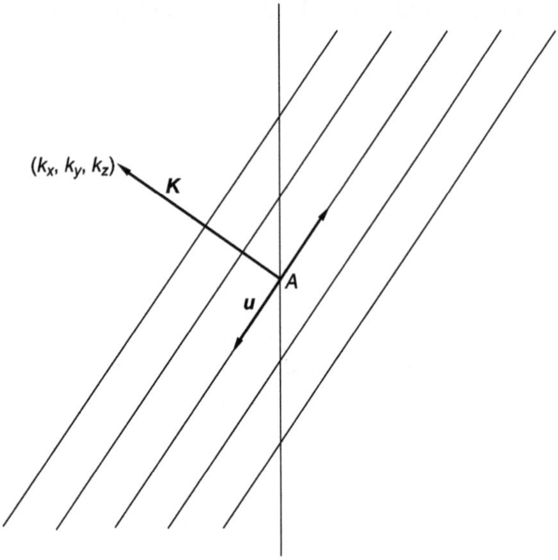

Fig. 1.4 Orientation of the wave and fluid velocity vectors

or

$$\mathbf{K} \cdot \mathbf{u} = 0 \tag{1.59}$$

From Eq. (1.59) it follows that the fluid velocity is perpendicular to the wave vector $\mathbf{K}(k_x, k_y, k_z)$ and it is directed along the crests of the waves, being perpendicular to the direction of phase propagation (see for example particle A in Fig. 1.4).

Following Pedlosky (2003) we define now the three-dimensional group velocity vector $\mathbf{C}_g$ as follows:

$$\mathbf{C}_g = C_{gx}\mathbf{i} + C_{gy}\mathbf{j} + C_{gz}\mathbf{k} = \frac{\partial \omega}{\partial k_x}\mathbf{i} + \frac{\partial \omega}{\partial k_y}\mathbf{j} + \frac{\partial \omega}{\partial k_z}\mathbf{k} \tag{1.60}$$

Using definition of frequency ω as in Eq. (1.52) yields:

$$\left.\begin{aligned} \frac{\partial \omega}{\partial k_x} &= \frac{N(z)}{K}\frac{k_z^2}{K^2}\frac{k_x}{k_h} \\ \frac{\partial \omega}{\partial k_y} &= \frac{N(z)}{K}\frac{k_z^2}{K^2}\frac{k_y}{k_h} \\ \frac{\partial \omega}{\partial k_z} &= -N(z)\frac{k_h k_z}{K^3} \end{aligned}\right\} \tag{1.61}$$

Therefore we have (Pedlosky 2003):

$$\mathbf{K} \cdot \mathbf{C}_g = k_x C_{gx} + k_y C_{gy} + k_z C_{gz} = \frac{N(z) k_z^2}{\omega K^4}(k_x^2 + k_y^2 - k_h^2) = 0 \tag{1.62}$$

The group velocity is perpendicular to the wave vector and parallel to the fluid velocity. The horizontal components of both velocities are in the same direction, while their vertical components are opposite. It means that the phase velocity has an upward component when the group velocity has a downward component, and *vice versa*. Therefore, the energy travels along the crests and troughs and not perpendicular to them. Also the wave energy flux $\mathbf{F} = \overline{p\mathbf{u}} = \mathbf{C}_g E$ is normal to $\mathbf{K}$.

This is in clear contrast to the surface waves, as for the internal waves, the group velocity is not only equal to the phase velocity, but also it propagates at a right angle to phase speed. The presence of a non-zero vertical component of group velocity implies that these waves transport wave energy upwards or downwards through the water column.

It should be noted that the magnitude of group velocity C_g can be written as (Thorpe 2005):

$$C_g = \cos(\theta)\sin(\theta)(N^2 - f^2)/(K\omega) \tag{1.63}$$

where angle θ is defined by Eq. (1.53).

In general, when we allow the waves to propagate in a three-dimensional space, under the influence of the Earth rotation, from Eq. (1.48) we have:

$$k_z^2 = \frac{N_0^2 - \omega^2}{\omega^2 - f^2} k_h^2 \tag{1.64}$$

where $N(z) \approx N_0$. However if we orient the x-axis in the direction of wave propagation, we obtain:

$$k_z^2 = \frac{N_0^2 - \omega^2}{\omega^2 - f^2} k_x^2 \tag{1.65}$$

Finally we note, that although solutions for internal waves are valid for a wide range of frequencies (see Eq. 1.51), some limiting cases are possible:

- *High-frequency range* ($\omega \approx N, but\, \omega \gg f$)
  In this range f^2 is negligible, the dispersion relation (1.65) simplifies as follows:

$$k_z^2 \approx \frac{k_x^2(N_0^2 - \omega^2)}{\omega^2} \quad \text{or} \quad \omega^2 \approx \frac{N_0^2 k_x^2}{k_x^2 + k_z^2} \tag{1.66}$$

  Using $\tan\theta = k_z/k_x$, we obtain $\omega = N_0 \cos\theta$, when θ is the angle between the wavenumber and the horizontal axis. Thus high-frequency waves are the same as the non-rotating internal waves.

- *Low-frequency range* ($\omega \ll N, but\, \omega \geq f$)
  In this range ω^2 can be neglected in comparison with N^2, and the dispersion relation (1.65) becomes:

$$k_z^2 \approx \frac{k_x^2 N_0^2}{\omega^2 - f^2} \qquad \text{or} \qquad \omega^2 \approx f^2 + N_0^2 \frac{k_x^2}{k_z^2} \tag{1.67}$$

  It should be noted that the low-frequency limit results from the hydrostatic assumption and neglecting the wave acceleration in the vertical equation of motion.

- *Mid-frequency range* ($f \ll \omega \ll N$)
  In this range, the dispersion relation (1.65) simplifies as follows:

$$k_z^2 \approx \frac{k_x^2 N_0^2}{\omega^2} \tag{1.68}$$

  In the approximation above, both the hydrostatic and non-rotating assumptions are applied.

We extend now above results to include, except the vertical stratification, represented by $N^2(z)$, the background horizontal flow $U(z)$. Thus, in the Eqs. (1.31)–(1.35) we replace velocity u with $U + u$ and assume small fluctuations in the vertical plane (x, z) and non-rotational motion. Therefore we have:

$$\frac{\partial u}{\partial t} + U\frac{\partial u}{\partial x} + w\frac{dU}{dz} = -\frac{1}{\overline{\rho}}\frac{\partial p}{\partial x} \tag{1.69}$$

$$\frac{\partial w}{\partial t} + U\frac{\partial w}{\partial x} = -\frac{1}{\overline{\rho}}\frac{\partial p}{\partial z} - g\frac{\rho}{\overline{\rho}} \tag{1.70}$$

$$\frac{\partial \rho}{\partial t} + U\frac{\partial \rho}{\partial x} = w\frac{\overline{\rho}}{g}N^2(z) \tag{1.71}$$

$$\frac{\partial u}{\partial x} + \frac{\partial w}{\partial z} = 0 \tag{1.72}$$

Following Sutherland (2010) we combine these equations in a single partial differential equation for the stream function ψ such as:

$$(u, w) = \left(-\frac{\partial \psi}{\partial x}, \frac{\partial \psi}{\partial z}\right) \tag{1.73}$$

For the stream function amplitude $\hat{\psi}$ we obtain:

$$\frac{d^2\hat{\psi}}{dz^2} + \left[\frac{N^2}{(U - C)^2} - \frac{U''}{(U - C)} - k_x^2\right]\hat{\psi} = 0 \tag{1.74}$$

where $C = \omega/k_x$ is the horizontal phase speed and $U'' = d^2U/dz^2$.

Equation (1.74) is known as the Taylor-Goldstein equation. This equation is singular in the sense that the coefficients of equation become infinitely large at some level $z = z_c$ (critical level) where horizontal phase speed $C = U(z_c)$. A Frobenius expansion exhibits the behaviour of solutions near critical level z_c. The structure of stream function amplitude $\hat{\psi}$ changes with the gradient Richardson number $Ri = N^2/(U')^2$ in the vicinity of critical level. It was found that the stratified shear flow remains stable when $Ri > 1/4$ for all z. Thus, a necessary condition for instability is that the gradient Richardson number is less than $1/4$ for some z.

It should be noted that the Taylor-Goldstein equation (1.74) can be used to predict the structure and dispersion relation of internal waves in a stratified shear flow for stable and unstable situations. For prescribed velocity U and frequency N, the solutions of Eq. (1.74) can be found numerically by integrating for given k_x and a guess at the possible complex eigenvalues $C(k_x)$.

Let us consider the solution of the Taylor-Goldstein equation (1.74) for a simple case of uniform flow $U(z) = U_0$ and uniformly stratified fluid, $N^2(z) = N_0^2$. The two solutions for amplitudes $\hat{\psi}$ are (Sutherland 2010):

$$\hat{\psi}(z) \sim C_+ \exp(i\gamma z) \quad \text{and} \quad \hat{\psi}(z) \sim C_- \exp(i\gamma z) \tag{1.75}$$

in which:

$$\gamma = k_x \sqrt{\frac{N_0^2}{\Omega_0^2} - 1} \tag{1.76}$$

Introducing a constant background flow induces the Doppler-shift of the frequency from its intrinsic value ω, measured by a stationary observer, to its extrinsic value $\Omega_0 = \omega - U_0 k_x$, measured with respect to the background flow. More discussion on the relations between frequencies ω and Ω_0 are given in Sect. 4.4. If the right-hand side of Eq. (1.76) is non-negative, we can consider γ as the vertical wavenumber k_z. Therefore we have:

$$\Omega_0^2 = (\omega - U_0 k_x)^2 = N_0^2 \frac{k_x^2}{k_x^2 + k_z^2} \tag{1.77}$$

The Taylor-Goldstein equation indicates that flow supports both upward- and downward-propagating plane waves. From Eq. (1.65) it follows that the vertical component of the group velocity is opposite in sign to the vertical component of the phase velocity, $k_z = +\gamma$ corresponds to downward propagating waves and $k_z = -\gamma$ corresponds to upward propagating waves. The more complicated cases of the shear stratified flow are considered by Sutherland (2010).

References

Apel JR, Holbrook JR, Liu AK, Tsai JJ (1985) The Sulu Sea internal soliton experiment. J Phys Oceanogr 15:1625–1651

Bockel M (1962) Traveaux océanographiques dé 'Origny' a Gibraltar. Cahiers Oéonographique 14:325–329

Botsford LW, Moloney CL, Hastings A, Largier JL, Powell TM, Higgins K, Quinn JF (1994) The influence of spatially and temporally varying oceanographic conditions on meroplanktonic metapopulations. Deep-Sea Res II 41:107–145

Bryden HL, Candela J, Kinder TH (1994) Exchange through the strait of Gibraltar. Prog Oceanogr 33:201–248

Burrage D, Massel SR, Steinberg C, Skirving W (1996) Detecting surface and internal wave signatures on the North-West shelf of Australia using the ERS-1 and ERS-2 Active Microwave Instruments (AMI). In: Proceedings of the First Australian ERS Symposium. CSIRO, Canberra, pp 11–26

Chong JC, Sprintall J, Hautala S, Morawitz WL, Bray NA, Pandre W (2000) Shallow throughflow variability in the outflow straits of Indonesia. Geophys Res Lett 27(1):125–128

Derzho OG, Grimshaw R (1997) Solitary waves with a vortex core in a shallow layer of stratified fluid. Phys Fluids 9:3378–3385

Dwi Susanto R, Gordon AL, Sprintall J, Herunadi B (2000) Interseasonal variability and tides in Makassar Strait. Geophys Res Lett 27(10):1499–1502

Dwi Susanto R, Mitnik L, Zheng Q (2005) Ocean internal waves observed in the Lombok Strait. Oceanography 18(4):81–87

Garrett C, Munk WH (1975) Space-time scales of internal waves: a progress report. J Geoph Res 80:291–313

Garrett C, Munk WH (1979) Internal waves in the ocean. Ann Rev Fluid Mech 11:339–369

Halpern D (1971) Observations of short-period internal waves in Massachusetts Bay. J Mar Res 29:116–132

Holloway PE (1983) Internal tides on the Australian North-West Shelf: a preliminary investigation. J Phys Oceanogr 14:1778–1790

Holloway PE (1984) On the semi-diurnal internal tide at a shelf-break region on the Australian North West Shelf. J Phys Oceanogr 14:1778–1790

Holloway PE (1994) Observations of internal tide propagation on the Australian North West Shelf. J Phys Oceanogr 24:1706–1716

Holloway PE (1996) A numerical model of internal tides with application to the Australian North West Shelf. J Phys Oceanogr 26:21–37

Huang NE, Shen SS (eds) (2005) Hilbert-Huang transform and its applications. World Scientific Publ., Singapore, 311 p

Huthnance JM (1989) Internal tides and waves near the continental shelf edge. Geophys Astro Fluid 48:81–106

Kahru M (1983) Phytoplankton patchiness generated by long internal waves: a model. Mar Ecol-Prog Ser 10:111–117

Konyaev KV, Sabinin KD, Serebryany A (1995) Large amplitude internal waves at the Mascarene Ridge in the Indian Ocean. Deep-Sea Res 42:2075–2091

Korteweg DJ, de Vries G (1895) On the change of form of long waves advancing in a rectangular canal, and on a new type of long stationary waves. Phil Mag J Sci 39:422–443

Kozlov I, Romanenkov D, Zimin A, Chapron B (2014) SAR observing large-scale nonlinear internal waves in the White Sea. Remote Sens Environ 147:99–107

Krauss W (1966) Interne Wellen. Gebruder Borntraeger, Berlin, 248 p

Leichter JJ, Shellenbarger G, Genovese SJ, Wing SR (1998) Breaking internal waves on a Florida (USA) coral reef: a plankton pump at work? Mar Ecol-Prog Ser 166:83–97

Lien RC, Henyey F, Ma B (2014) Large-amplitude internal solitary waves observed in the Northern South China Sea: properties and energetics. J Phys Oceanogr 44(4):1095–1115

Lucas AJ, Franks PJS, Dupont CL (2011) Horizontal internal-tide fluxes support elevated phytoplankton productivity over inner continental shelf. Limnol Oceanogr: Fluids Environ 1(1):56–74
Manasseh R, Chin CY, Fernando HI (1998) The transition from density-driven to wave-dominated isolated flows. J Fluid Mech 361:253–274
Massel SR (1999) Fluid mechanics for marine ecologists. Springer, Berlin, 566 p
Massel SR (2013) Ocean surface waves; their physics and prediction, 2nd edn. World Scientific Publ, Singapore, 676 p
Munk WH (1966) Abyssal recipes. Deep-Sea Res Oceanogr Abs 13(4):707–730
New AL, Pingree RD (1990) Large amplitude soliton packets in the central Bay of Biscay. Deep-Sea Res 37A:513–524
Osborne AR, Burch TL (1980) Internal solutions in the Andaman Sea. Science 208(4443):451–460
Pedlosky J (2003) Wave in the ocean and atmosphere. Introduction to wave dynamics. Springer, Berlin, 260 p
Phillips OM (1977) The dynamics of the upper ocean, 2nd edn. Cambridge University Press, Cambridge, 336 p
Roberts J (1975) Internal gravity waves in the ocean. Marcel Dekker Inc., New York, 274 p
Russell JS (1844) Report to committee on waves. In: Report of the 14th meeting of the british association of the advancement of science. British Association of the Advancement of Science, York, pp 311–390
Sabinin KD (1992) Internal wave packets over the Mascarene Ridge. Fizika Atm i Okeana 28:625–633 (in Russian)
Serebryanyj AN (1990) Effects of nonlinearities on the internal waves on shelfs. Fizika Atm i Okeana 26:285–293 (in Russian)
Sutherland B (2010) Internal gravity waves. Cambridge University Press, Cambridge, 377 p
Thorpe SA (1968) On the shape of progressive internal waves. Philos T Roy Soc A 263:563–614
Thorpe SA (2005) The turbulent ocean. Cambridge University Press, Cambridge, 484 p
UNESCO (1981) Tenth report of the joint panel on oceanographic tabels and standards. UNESCO Tech Pap Mar Sci, Paris, 25 p
Van Gastel P, Ivey GN, Meuleners MJ, Antenucci JP, Fringer O (2009) The variability of the large-amplitude internal wave field on the Australian North West Shelf. Cont Shelf Res 29:1373–1383
Zheng Q, Klemas V, Yan XH, Pan J (2001) Nonlinear evolution of ocean internal solitons propagating along an inhomogeneous thermocline. J Geophys Res 106(C7):14083–14094

Chapter 2
Internal Waves in Two-Layer Sea

2.1 Introduction

Two-layer model of the vertical structure of ocean water, where a layer of water of uniform lower density is located over a layer of uniform higher density, with a sharp interface, is quite realistic for coastal regions. River runoff occupies the upper layer of low density, over the deep layer of much higher salinity, and gradient of salinity (i.e. density) occurs between them. Two-layer model is also applicable when one tries to describe an upper, well-mixed oceanic layer located over deeper water. The thermocline is then fairly abrupt and separates water masses above and below, each of which is almost homogeneous. This scenario is sometimes observed during influx of the saline water from the North Sea to the Baltic Sea (Piechura and Beszczyńska-Möller 2004).

The density gradient layer is unstable and under some impulse, disturbances of the waving form are generated. These disturbances are known as the interfacial internal waves. In the following sections two types of solutions for the interfacial waves are discussed. In Sect. 2.2, it is shown that for the infinitesimal interfacial waves, a simple analytical solutions are possible. The final forms for the wave kinematics parameters and vertical structure of wave velocities have been developed. If the amplitude of interfacial waves is higher, the nonlinear effects become visible. For deep and intermediate waters, the solution is obtained in Sect. 2.3 by applying the Stokes's type expansion in the wave steepness. In the same section, the stable solution of the very well known Korteweg-de Vries equation for long interfacial waves is given. In particular, the solitary and cnoidal type solution are discussed in detail.

2.2 Analytic Solutions for Infinitesimal Interfacial Waves

We consider a simple case of a two-layer ocean when the height over which the density change between the layers is small in comparison with the wavelength. Therefore the termocline can be represented approximately by infinitesimally thin interface.

S.R. Massel, *Internal Gravity Waves in the Shallow Seas*,
GeoPlanet: Earth and Planetary Sciences, DOI 10.1007/978-3-319-18908-6_2

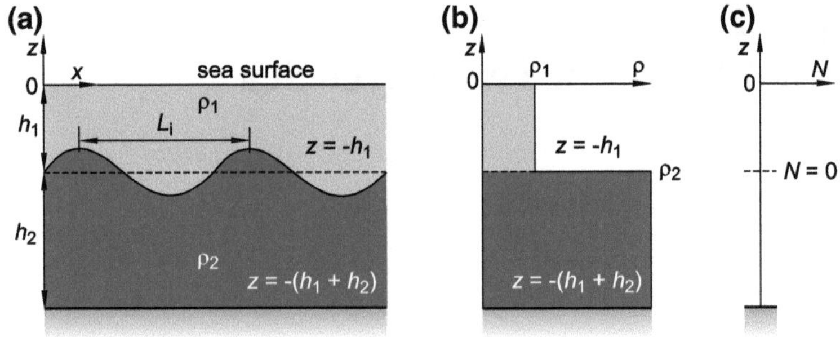

Fig. 2.1 Two-layer ocean model: **a** two layers of fluid of different densities, **b** vertical density profile, **c** vertical profile of frequency (Massel 1999)

We assume that an upper layer of depth h_1 and density ρ_1 is lying on a lower layer of depth h_2 and density ρ_2. The total water depth is equal $h = h_1 + h_2$. An internal wave of frequency ω is propagating at the interface between the layers (see Fig. 2.1). It should be noted that at the level $z = -h_1$, the gradient of water density becomes infinitely large. Thus the Väisälä-Brunt frequency, $N(z)$, cannot be defined at that level, and in both layers, the frequency $N(z)$ is simply equal zero.

Let us assume that $\omega^2 \gg f^2$ and therefore the Earth rotation is neglected. This condition is usually satisfied for relatively short internal waves. For example, for station located at the latitude $\varphi \sim 50°$, and for internal waves period $1/4\,\mathrm{h} < \mathrm{T} < 3\,\mathrm{h}$, the ratio of the inertial frequency f and internal wave frequency ω is $0.016 < (f/\omega) < 0.191$.

If the fluid is irrotational and incompressible in both layers, the governing equations describing the wave motion can be presented in the form of velocity potentials (Sutherland 2010):

$$\left.\begin{aligned} \nabla^2\phi_1 &= 0 \text{ for } -h_1 \le z \le 0 \\ \nabla^2\phi_2 &= 0 \text{ for } -h \le z \le -h_1 \end{aligned}\right\} \tag{2.1}$$

with the following kinematic and dynamic boundary conditions:

$$\frac{\partial\phi_1}{\partial z} = \frac{\partial\phi_2}{\partial z} = \frac{\partial\zeta}{\partial t} \quad \text{at} \quad z = -h_1 \tag{2.2}$$

and

$$\rho_1\frac{\partial\phi_1}{\partial t} + \rho_1 g\zeta = \rho_2\frac{\partial\phi_2}{\partial t} + \rho_2 g\zeta \quad \text{at} \quad z = -h_1 \tag{2.3}$$

$$\frac{\partial\phi_1}{\partial z} = 0 \quad \text{at} \quad z = 0 \tag{2.4}$$

$$\frac{\partial \phi_2}{\partial z} = 0 \quad \text{at} \quad z = -h \tag{2.5}$$

with the interfacial displacement ζ assumed in the form $\zeta(x, t) = A\cos(kx - \omega t)$.

We adopt the velocity potentials ϕ_1 and ϕ_2 to the boundary value problem as follows:

$$\phi_1(x, z, t) = -\frac{\omega}{k} A \frac{\cosh(kz)}{\sinh(kh_1)} \sin(kx - \omega t) \tag{2.6}$$

$$\phi_2(x, z, t) = \frac{\omega}{k} A \frac{\cosh[k(z + h)]}{\sinh(kh_2)} \sin(kx - \omega t) \tag{2.7}$$

Substituting Eqs. (2.6) and (2.7) into Eq. (2.3), we get the dispersion relation and the phase velocity C_0 as follows:

$$\omega^2 = \frac{gk(\rho_2 - \rho_1)}{\rho_1 \coth(kh_1) + \rho_2 \coth(kh_2)} \tag{2.8}$$

and

$$C_0 = \frac{\omega}{k} = \sqrt{\frac{g}{k} \frac{(\rho_2 - \rho_1)}{\rho_1 \coth(kh_1) + \rho_2 \coth(kh_2)}} \tag{2.9}$$

The fluid velocities u and w in both layers can be found from the velocity potentials, i.e.:

– upper layer

$$\left.\begin{aligned} u_1(x, z, t) &= \frac{\partial \phi_1}{\partial x} = -\omega A \frac{\cosh(kz)}{\sinh(kh_1)} \cos(kx - \omega t) = U_1(z)\cos(kx - \omega t) \\ w_1(x, z, t) &= \frac{\partial \phi_1}{\partial z} = -\omega A \frac{\sinh(kz)}{\sinh(kh_1)} \sin(kx - \omega t) = W_1(z)\sin(kx - \omega t) \end{aligned}\right\} \tag{2.10}$$

– lower layer

$$\left.\begin{aligned} u_2(x, z, t) &= \frac{\partial \phi_2}{\partial x} = \omega A \frac{\cosh[k(z + h)]}{\sinh(kh_2)} \cos(kx - \omega t) = U_2(z)\cos(kx - \omega t) \\ w_2(x, z, t) &= \frac{\partial \phi_2}{\partial z} = \omega A \frac{\sinh[k(z + h)]}{\sinh(kh_2)} \sin(kx - \omega t) = W_2(z)\sin(kx - \omega t) \end{aligned}\right\} \tag{2.11}$$

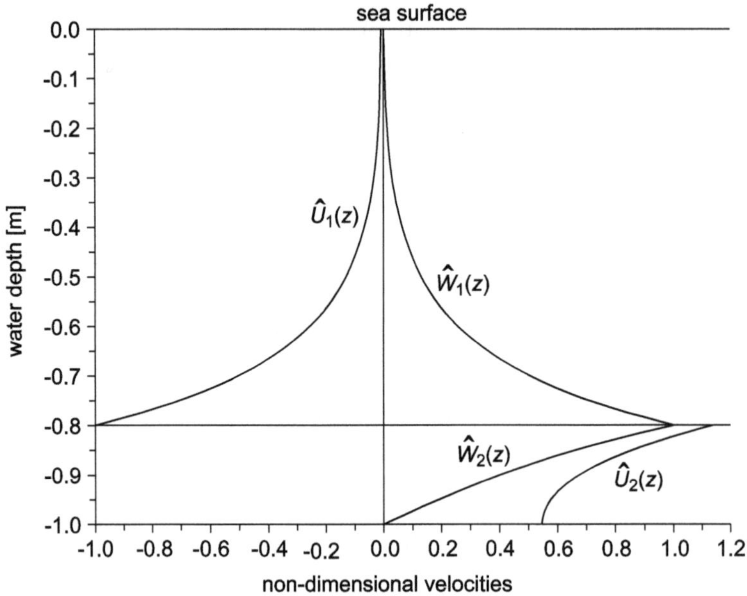

Fig. 2.2 Vertical profiles of non-dimensional velocity amplitudes in two-layer fluid (deep water case)

For illustration, we consider a laboratory experimental case when the total water depth $h = 1.0\,\text{m}$. The water depths and densities in upper and lower layers are equal $h_1 = 0.8\,\text{m}$, $\rho_1 = 1005\,\text{kg}\,\text{m}^{-3}$, and $h_2 = 0.2\,\text{m}$, $\rho_2 = 1025\,\text{kg}\,\text{m}^{-3}$, respectively. Interfacial wave period is equal $T = 8\,\text{s}$. From Eqs. (2.8) and (2.9) it follows that wavelength $L = 0.919\,\text{m}$ and the phase velocity $C_0 = 0.115\,\text{m}$.

In Fig. 2.2 the vertical profiles of the non-dimensional velocity amplitudes ($\hat{U}(z) = U(z)/\omega A$ and $\hat{W}(z) = W(z)/\omega A$) are shown. The vertical and horizontal velocity amplitudes decay from the interface line to the sea surface and sea bottom. The horizontal velocity amplitude in the lower layer is opposite to that in the upper layer. Please note that in the Fig. 2.2, the horizontal velocity at $z = 0$ does not exactly equal zero but in our case it is equal to $\hat{U}_1(z) = -0.00844$. Let us reconstruct an approximate path followed by an individual fluid parcel during the passage of a small-amplitude internal wave. We assume that parcel motion around a fixed point (x_0, z_0) is small. Therefore, the actual position of parcel $(x(t), z(t))$ at time t is given by:

–upper layer

$$\left.\begin{aligned} x(t) &\approx x_0^{(1)} + \int_0^t u_1 dt \\ z(t) &\approx z_0^{(1)} + \int_0^t w_1 dt \end{aligned}\right\} \tag{2.12}$$

–lower layer

$$\left.\begin{aligned} x(t) &\approx x_0^{(2)} + \int_0^t u_2 dt \\ z(t) &\approx z_0^{(2)} + \int_0^t w_2 dt \end{aligned}\right\} \tag{2.13}$$

Substituting Eqs. (2.10) and (2.11) into Eqs. (2.12) and (2.13) we obtain:
–upper layer

$$\left.\begin{aligned} x(t) &\approx x_0^{(1)} - \frac{U_1\left(z_0^{(1)}\right)}{\omega} \sin\left(kx_0^{(1)} - \omega t\right) \\ z(t) &\approx z_0^{(1)} + \frac{W_1\left(z_0^{(1)}\right)}{\omega} \cos\left(kx_0^{(1)} - \omega t\right) \end{aligned}\right\} \tag{2.14}$$

–lower layer

$$\left.\begin{aligned} x(t) &\approx x_0^{(2)} - \frac{U_2\left(z_0^{(2)}\right)}{\omega} \sin\left(kx_0^{(2)} - \omega t\right) \\ z(t) &\approx z_0^{(2)} + \frac{W_2\left(z_0^{(2)}\right)}{\omega} \cos\left(kx_0^{(2)} - \omega t\right) \end{aligned}\right\} \tag{2.15}$$

Under the assumption of small-amplitude wave, the displacements from the original position are small, so we can approximate $x \sim x_0$ and $z \sim z_0$ on the right-hand sides of Eqs. (2.14) and (2.15).

Eliminating time t in these equations, we obtain the elliptic path of the parcel in the form:

$$\frac{\left(x - x_0^{(1)}\right)^2}{A_1^2} + \frac{\left(z - z_0^{(1)}\right)^2}{B_1^2} = 1 \tag{2.16}$$

and

$$\frac{\left(x - x_0^{(2)}\right)^2}{A_2^2} + \frac{\left(z - z_0^{(2)}\right)^2}{B_2^2} = 1 \tag{2.17}$$

in which:

$$\frac{A_1}{A} = \frac{\cosh\left(kz_0^{(1)}\right)}{\sinh(kh_1)}, \quad \frac{B_1}{A} = \frac{\sinh\left(kz_0^{(1)}\right)}{\sinh(kh_1)} \tag{2.18}$$

$$\frac{A_2}{A} = \frac{\cosh\left[k\left(z_0^{(2)} + h\right)\right]}{\sinh(kh_2)}, \quad \frac{B_2}{A} = \frac{\sinh\left[k\left(z_0^{(2)} + h\right)\right]}{\sinh(kh_2)} \tag{2.19}$$

The orbit dimensions close to the interface are:

$$\frac{A_1}{A} = \frac{A_2}{A} \sim \coth(kh_1), \quad \frac{B_1}{A} = \frac{B_2}{A} \sim 1.0 \tag{2.20}$$

and they decay from the interface; namely, close to the sea surface we have:

$$\frac{A_1}{A} \sim \frac{1}{\sinh(kh_1)} \quad \text{and} \quad \frac{B_1}{A} \sim 0 \tag{2.21}$$

while close to sea bottom they become:

$$\frac{A_2}{A} \sim \frac{1}{\sinh(kh_2)} \quad \text{and} \quad \frac{B_2}{A} \sim 0 \tag{2.22}$$

All formulae for wave velocity amplitudes and dispersion relation for some combinations of the water depths h_1 and h_2 can be simplified. In particular, for the deep water waves when $kh_1 \gg 1$ and $kh_2 \gg 1$, the dispersion relation and phase velocity are as follows:

$$\omega = \sqrt{gk\frac{\rho_2 - \rho_1}{\rho_2 + \rho_1}} = \varepsilon\sqrt{gk}, \quad C_0 = \varepsilon\sqrt{\frac{g}{k}} \tag{2.23}$$

where:

$$\varepsilon = \sqrt{\frac{\rho_2 - \rho_1}{\rho_2 + \rho_1}} \tag{2.24}$$

The quantity ε is a small number if density difference between the two layers is small. From Eq. (2.23) it follows that waves at the interface between two liquids of infinite thickness travel like deep water surface waves, but at a much reduced frequency due to small parameter ε. Consequently, the phase speed will be much smaller than speed of surface waves. For example, if $\rho_1 = 1020\,\text{kg m}^{-3}$ and $\rho_2 = 1023\,\text{kg m}^{-3}$, Eq. (2.23) gives:

$$C_0 = 0.038\sqrt{\frac{g}{k}} = 0.038C \tag{2.25}$$

where C is the phase velocity of the surface wave of this length.

On the other hand, for shallow water long waves, when $kh_1 \ll 1$ and $kh_2 \ll 1$, the phase velocity becomes:

$$C_0 = \sqrt{\frac{g(\rho_2 - \rho_1)\, h_1 h_2}{\rho_2 h_1 + \rho_1 h_2}} \tag{2.26}$$

Let us assume that densities ρ_1 and ρ_2 are the same as above, and additionally $h_1 = h_2$. Thus, Eq. (2.26) gives:

$$C_0 = \varepsilon\sqrt{gh_2} = \varepsilon C = 0.038C \tag{2.27}$$

Again the phase speed of an internal wave is much smaller than that of a long surface wave of this length. These waves are non-dispersive and travel slowly because of small density difference.

In the coastal regions, near river estuaries, the upper layer is usually very thin, say $kh_1 \ll 1$, over a deep lower layer, where usually $kh_2 \gg 1$. Using these water depth conditions in Eq. (2.9), we obtain:

$$C_0 \approx \sqrt{\frac{\rho_2 - \rho_1}{\rho_1}\, gh_1} \tag{2.28}$$

For example, if $h_1 = 5\,\mathrm{m}$, $\rho_1 = 1000\,\mathrm{kg\,m^{-3}}$ (fresh water), and $\rho_2 = 1023\,\mathrm{kg\,m^{-3}}$ (salt water), the phase velocity, C_0, becomes $0.15\sqrt{gh_1} = 1.05\,\mathrm{m\,s^{-1}}$.

For shallow water, when $kh_1 \ll 1$ and $kh_2 \ll 1$, both components of the internal wave velocities become approximately linear functions of z coordinate. From Eqs. (2.10) and (2.11) we obtain the amplitudes of velocities in the form:

– upper layer

$$\left.\begin{aligned} U_1(z) &= -\omega A\,\frac{\cosh(kz)}{\sinh(kh_1)} \approx -\frac{\omega A}{kh_1} = -\frac{AC_0}{h_1} \\ W_1(z) &= -\omega A\,\frac{\sinh(kz)}{\sinh(kh_1)} \approx -\omega A\frac{z}{h_1} = -\frac{AC_0}{h_1}\,(kz) \end{aligned}\right\} \tag{2.29}$$

– lower layer

$$\left.\begin{aligned} U_2(z) &= \omega A\,\frac{\cosh[k(z+h)]}{\sinh(kh_2)} \approx \frac{\omega A}{kh_2} = \frac{AC_0}{h_2} \\ W_2(z) &= \omega A\,\frac{\sinh[k(z+h)]}{\sinh(kh_2)} \approx \frac{\omega A}{h_2}(z+h) = \frac{AC_0}{h_2}\,k(z+h) \end{aligned}\right\} \tag{2.30}$$

In Fig. 2.3 the vertical profiles of the non-dimensional amplitudes $\hat{U}(z) = U(z)/\omega A$ and $\hat{W}(z) = W(z)/\omega A$ are illustrated for the case of shallow water. The interface is

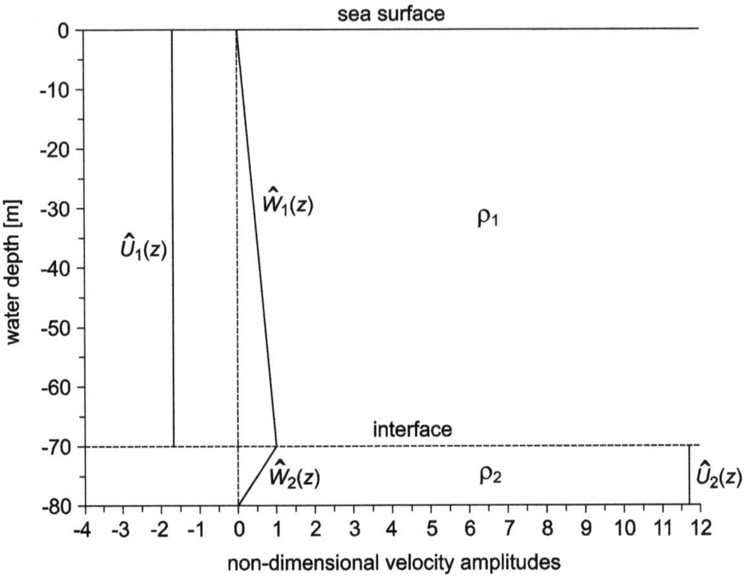

Fig. 2.3 Vertical profiles of non-dimensional velocity amplitudes in two-layer fluid (shallow water case)

located at the water depth $h_1 = 70\,\mathrm{m}$ and the densities of the upper and lower layers are $1004\,\mathrm{kg\,m^{-3}}$ and $1012\,\mathrm{kg\,m^{-3}}$, respectively. The internal wave of amplitude A, period $T = 15\,\mathrm{min}$ ($\omega = 0.00698\,\mathrm{rad\,s^{-1}}$) and wavelength $L = 736\,\mathrm{m}$ ($k = 0.00854\,\mathrm{m^{-1}}$) propagates along the interface with phase velocity $C_0 = 0.818\,\mathrm{m\,s^{-1}}$. The vertical velocity amplitude is a linear function of z with discontinuous gradient at the interface line. The horizontal velocity amplitude does not decay with depth and in the lower layer it is opposite to that in the upper layer.

We also note that internal waves in the ocean produce only a very small vertical displacements of free surface; usually they are smaller by a factor of the order $(\rho_2 - \rho_1)/\rho_1$ (say $\sim 10^{-3}$) than the internal wave displacement.

Suppose now that an object is located at water depth $z = -h_1$ (see Fig. 2.1). This object can be stationary when the balance between its weight and the buoyancy force is maintained. However, due to propagation of the interfacial waves, the object's buoyancy varies depending on the density of the surrounding fluid. In his book Pinet (1992) remarked the nuclear submarine *U.S.S. Thresher*, which was lost in the West Atlantic in 1963 with all crew members. There had been no indication of equipment malfunction or unusual storm weather. As submerged submarines attain neutral buoyancy by flooding or jettisoning sea water from ballast tanks, there are speculations that *U.S.S. Thresher* was probably cruising along a pycnocline when it encountered a large internal wave and suddenly dropped to a greater depth because of the lower density of the ambient water. Evidently the incident occurred too rapidly and the submarine crew was not able to arrest the ship's fall.

2.3 Progressive Interfacial Waves of Finite Amplitude

2.3.1 Deep and Intermediate Waters

Solutions discussed in the previous section are valid under the assumption of infinitesimal waves when the higher order terms are neglected. Here we consider waves of permanent form and finite amplitude propagating at the interface between two fluids of finite depths. Let the wavelength, L, be not too much greater than the fluid depth. For such case, Hunt (1961) provided the third order solution for two-fluid system of finite depths and bounded by upper and bottom rigid horizontal planes. The shape of the interfacial wave is:

$$\zeta(x,t) = A_1 \cos(kx - \omega t) + A_2 \cos 2(kx - \omega t) + \\ + A_3 \cos 3(kx - \omega t) \tag{2.31}$$

where A_1, A_2 and A_3 are the amplitudes of components of first, second and third order, respectively. The symbols k and ω denote the wavenumber and wave frequency.

For amplitudes A_2 and A_3 we have (Thorpe 1968):

$$A_2 = \frac{A_1(kA_1)}{4}\left[\rho_2 \frac{(3-T_2)^2}{T_2^2} - \rho_1 \frac{(3-T_1)^2}{T_1^2}\right](\rho_1 T_1 + \rho_2 T_2)^{-1} \tag{2.32}$$

and

$$A_3 = \frac{kA_1}{16}\left(S_{31}A_2 - S_{32}(kA_1)A_1\right) S_{33}^{-1} \tag{2.33}$$

$$S_{31} = \frac{9 + 4T_2^2 + 3T_2^4}{T_2^2(3+T_2^2)}\rho_2 - \frac{9 + 4T_1^2 + 3T_1^4}{T_2^2(3+T_1^2)}\rho_1 \tag{2.34}$$

$$S_{32} = \frac{3 - T_2^2}{T_2^2(3+T_2^2)}\rho_2 + \frac{3 - T_1^2}{T_1^2(3+T_1^2)}\rho_1 \tag{2.35}$$

$$S_{33} = \frac{\rho_1 T_1}{3+T_1^2} + \frac{\rho_2 T_2}{3+T_2^2} \tag{2.36}$$

with:

$$T_1 = \tanh(kh_1), \quad T_2 = \tanh(kh_2) \tag{2.37}$$

The equations above are simplified for particular water depths. If depths of upper and lower layers are the same ($h_1 = h_2 = h$), the amplitudes A_2 and A_3 become as follows (Thorpe 1968):

$$A_2 = \frac{1}{4} A_1(kA_1) \frac{\rho_2 - \rho_1}{\rho_1 + \rho_2} \frac{3 - T^2}{T^3} \tag{2.38}$$

$$A_3 = \frac{1}{64} A_1(kA_1)^2 \frac{3 - T^2}{T^6} S_{34} \tag{2.39}$$

where:

$$S_{34} = \frac{(\rho_2 - \rho_1)^2 (9 + 4T^2 + 3T^4) - 4(\rho_1 + \rho_2)^2 T^2}{(\rho_1 + \rho_2)^2} \tag{2.40}$$

$$T = \tanh(kh) \tag{2.41}$$

The corresponding dispersion relation is:

$$\omega^2 = gkT \frac{\rho_2 - \rho_1}{\rho_1 + \rho_2} S_d \tag{2.42}$$

where:

$$S_d = 1 + \frac{(kA_1)^2}{8T^4(\rho_1 + \rho_2)} \left[4(\rho_1 + \rho_2)^2 T^2(2T^2 - 1) + (\rho_2 - \rho_1)^2 (3 - T^2)\right] \tag{2.43}$$

Now we remove the rigid upper boundary and express the fluid surface with accuracy to the second order as follows (Thorpe 1968):

$$\zeta_s(x, t) = a_1 \cos(kx - \omega t) + a_2 \cos 2(kx - \omega t) \tag{2.44}$$

while the equation of the interface displacement to the second order becomes:

$$\zeta(x, t) = A_1 \cos(kx - \omega t) + A_2 \cos 2(kx - \omega t) \tag{2.45}$$

The amplitudes A_1, A_2 and a_1, a_2 are given by Thorpe (1968), and the dispersion relation is:

$$\omega^4(\rho_1 T_1 T_2 + \rho_2) - \omega^2 gk\rho_2(T_1 + T_2) + (\rho_2 - \rho_1)T_1 T_2(gk)^2 = 0 \tag{2.46}$$

A sufficient condition for the motion of the upper surface to be neglected is:

$$\cosh(kh_1)\frac{\rho_2 - \rho_1}{\rho_2} \frac{\tanh(kh_1)}{\tanh(kh_1) + \tanh(kh_2)} \tag{2.47}$$

If the above inequality becomes true, we may suppose that the fluid has a fixed horizontal boundary as it was considered by Hunt (1961) (see Eq. 2.31).

A special type of distortion of the interface shape appears when kh_2 is large while kh_1 remains small, and the density difference is small. From Eqs. (2.31) and (2.32) we obtain (Thorpe 1968):

$$\zeta(x,t) = A_1 \cos(kx - \omega t) - \frac{3}{4} A_1(kA_1)\left(\frac{1-T_1}{T_1^2}\right)\cos 2(kx - \omega t) \qquad (2.48)$$

In this profile, the troughs are narrower than the crests, just opposite to the free surface waves in a homogeneous fluid.

If kh_1 is large and the density difference between the fluids is small again, the interface shape becomes:

$$\zeta(x,t) = A_1 \cos(kx - \omega t) + \frac{3}{4} A_1(kA_1)\left(\frac{1-T_2}{T_2^2}\right)\cos 2(kx - \omega t) \qquad (2.49)$$

which is similar to the surface gravity wave.

In general, when the fluid is bounded by a rigid surface above, the crests of wave profile may be narrower or wider than the troughs depending on whether $\rho_2(3 - T_2^2)/T_2^2$ is greater or less than $\rho_1(3 - T_1^2)/T_1^2$.

The extension of the second-order solution for the random internal waves propagating in a two-layer fluid was considered by Song (2004). He used a methodology of Sharma and Dean (1979) to discuss both the progressive and standing waves. When the upper surface of the two-layer fluid is free, the second-order solution for random internal waves and surface waves was developed by Liu (2006).

2.3.2 *Long Interfacial Waves*

2.3.2.1 General Remarks

If the wavelength of the internal waves is significantly longer than the characteristic fluid depth, the linear waves are non-dispersive with z-independent horizontal velocity and linear variation with z of the vertical velocity. However, when wave amplitude A is not too small, the weakly nonlinear effect produces wave different from the sinusoidal wave shape by wave crests steepening. Moreover, if the horizontal extent of waves is not too long, compared to the fluid depth h, the weak dispersion appears, broadening the waves. In most efforts to describe the evolution of these waves the Kortweg-de Vries (KdV) theory (Korteweg and de Vries 1895;

Miles 1981) is usually applied. In this theory, a balance between nonlinearity and dispersion of waves is parameterized by two nonlinear variables:

$$\alpha = \frac{A}{h}, \quad \beta = \left(\frac{h}{l}\right)^2 \tag{2.50}$$

where A is the wave amplitude and l is the wavelength scale. For this balance, both parameters must be of the same order of magnitude and both small, i.e. $\beta = O(\alpha) \ll 1$. If these two effects are balanced, solution of the Korteweg-de Vries equation results in steady waves such as solitary waves or cnoidal waves (Whitham 1974; Miles 1981; Massel 1989).

To exploit further the nonlinear and dispersive properties of the solitary and cnoidal internal waves, an extension of the Korteweg-de Vries (KdV) equation, known as the extended KdV (eKdV) equation or Gardner equation including cubic nonliearity is employed (Helfrich and Melville 1986, 2006; Grimshaw et al. 2007; Pelinovsky et al. 1994, 2007):

$$\frac{\partial \zeta}{\partial t} + (C_0 + \alpha\zeta + \alpha_1\zeta^2)\frac{\partial \zeta}{\partial x} + \beta\frac{\partial^3 \zeta}{\partial x^3} = 0 \tag{2.51}$$

in which ζ is the displacement of the interfacial wave and the linear phase velocity C_0 is the eigenvalue of the Sturm-Louville problem for the eigenmode. The coefficients α, α_1 and β are functions of the vertical stratification of fluid under the Boussinesq approximation.

To apply the extended KdV (eKdV) equation for the interfacial waves propagating in a two-layer fluid we define coefficients α, α_1 and β as follows (Helfrich and Melville 2006):

$$C_0 = \sqrt{\left(\frac{g\sigma h_1 h_2}{h_1 + h_2}\right)} \tag{2.52}$$

$$\alpha = \frac{3}{2}C_0\frac{h_1 - h_2}{h_1 h_2} \tag{2.53}$$

$$\alpha_1 = \frac{3C_0}{(h_1 h_2)^2}\left[\frac{7}{8}(h_1 - h_2)^2 - \left(\frac{h_1^3 + h_2^3}{h_1 + h_2}\right)\right] \tag{2.54}$$

$$\beta = \frac{C_0}{6}h_1 h_2 \tag{2.55}$$

where relative layer density difference $\sigma = 2(\rho_2 - \rho_1)/(\rho_1 + \rho_2) \ll 1$. The parameter β is always positive, the nonlinear parameter α however can be either positive or negative. If depth layer $h_1 = h_2$, and parameter $\alpha = 0$, it is necessary also to take into account the cubic nonlinearity. When $h_1 > h_2$, the pycnocline is displaced upward, in the opposite case ($h_2 > h_1$), the internal wave changes polarity and the depression of pycnocline is observed. This phenomenon was clearly illustrated by Lien et al. (2014) and Grimshaw et al. (2014) for internal solitary waves in the South China Sea. The effect of the change in sign of α on internal solitary waves leads to complex pictures of nonlinear transformation, including breaking of solitons and shocks on both the front and back faces of the waves. General relations for the coefficients with continuous stratification and shear fluid flow have been discussed by Grimshaw et al. (2004)—see also Chap. 3.

2.3.2.2 Solitary Waves

When α_1 is set to zero, Eq. (2.51) reduces to the classical KdV equation:

$$\frac{\partial \zeta}{\partial t} + C_0 \frac{\partial \zeta}{\partial x} + \alpha \zeta \frac{\partial \zeta}{\partial x} + \beta \frac{\partial^3 \zeta}{\partial x^3} = 0 \tag{2.56}$$

Let us find the stable solutions of this equation for the case when interfacial waves are propagating in a constant water depth. We first transform Eq. (2.51) to a frame moving at some constant speed U. Thus we can write $\zeta(x, t) = \zeta(X) = \zeta(x - Ut)$. After substituting this representation into Eq. (2.51), multiplying by $(\partial \zeta / \partial X)$ and integrating we obtain:

$$\left(\frac{\partial \zeta}{\partial X}\right)^2 + \frac{\hat{\alpha}}{3\hat{\beta}} \zeta^3 - \left(\frac{U}{C_0} - 1\right) \frac{1}{\hat{\beta}} \zeta^2 + \left(\frac{2}{C_0 \hat{\beta}} D_1\right) \zeta + \left(\frac{2}{C_0 \hat{\beta}} D_2\right) = 0 \tag{2.57}$$

in which D_1 and D_2 are integration constants, and

$$\hat{\alpha} = \frac{\alpha}{C_0}, \quad \hat{\beta} = \frac{\beta}{C_0} \tag{2.58}$$

The solitary wave, which is one of the possible solutions of the KdV equation, requires that displacement ζ and its derivatives tend to zero at infinity. Thus $D_1 = D_2 = 0$, and Eq. (2.57) takes the form (Whitham 1974):

$$\frac{3\hat{\beta}}{\hat{\alpha}} \left(\frac{d\zeta}{dX}\right)^2 = \zeta^2 (A - \zeta) \tag{2.59}$$

in which the amplitude A (maximum vertical displacement above the equilibrium fluid level) is:

$$A = \frac{3}{\hat{\alpha}} \left(\frac{U}{C_0} - 1\right) \tag{2.60}$$

and the solitary wave speed is:

$$U = C_0\left(1 + \frac{\hat{\alpha}}{3}A\right) = C_0\left(1 + \frac{1}{2}\frac{h_1^2 - h_2^2}{h_1 h}\frac{A}{h_2}\right) \tag{2.61}$$

From Eq. (2.59), the shape of the solitary wave becomes:

$$\zeta(x,t) = A\cosh^{-2}\left[\left(\frac{\hat{\alpha}A}{12\hat{\beta}}\right)^{1/2}(x - Ut)\right] \tag{2.62}$$

or

$$\zeta(x,t) = A\cosh^{-2}\left\{\sqrt{\frac{3A}{4hh_2^2}\left[1 - \left(\frac{h_2}{h_1}\right)^2\right]}(x - Ut)\right\} \tag{2.63}$$

Let us define the characteristic water depth h_r as follows:

$$h_r^3 = hh_2^2\left[1 - \left(\frac{h_2}{h_1}\right)^2\right]^{-1} \tag{2.64}$$

Thus we can represent the solitary internal wave in the classical form:

$$\zeta(x,t) = A\cosh^{-2}\left\{\sqrt{\frac{3A}{4h_r^3}}(x - Ut)\right\} \tag{2.65}$$

The horizontal velocities induced by the internal wave for both layers are equal, respectively (Zheng et al. 2001):

$$u_1(x,t) = -\frac{AC_0}{h_1}\cosh^{-2}\left\{\sqrt{\frac{3A}{4h_r^3}}(x - Ut)\right\} \tag{2.66}$$

and

$$u_2(x,t) = \frac{AC_0}{h_2}\cosh^{-2}\left\{\sqrt{\frac{3A}{4h_r^3}}(x - Ut)\right\} \tag{2.67}$$

It should be noted that the solution of the extended KdV equation (eKdV) Eq. (2.51) for solitary wave takes the form (Helfrich and Melville 2006):

$$\zeta(x,t) = \frac{A}{b + (1 - b)\cosh^2[\gamma(x - Ut)]} \tag{2.68}$$

in which:

$$U = C_0 + \frac{A}{3}\left(\alpha + \frac{1}{2}\alpha_1 A\right), \quad \gamma^2 = \frac{A\left(\alpha + \frac{1}{2}\alpha_1 A\right)}{12\beta}, \quad b = \frac{-A\alpha_1}{2\alpha + \alpha_1 A} \tag{2.69}$$

2.3.2.3 Cnoidal Waves

Except for the solitary wave, another permanent form of the long internal wave profile exists, being also a solution of the KdV equation. This solution is described by the elliptic function $cn(\theta; m)$ and it is known as the cnoidal wave. The length of the cnoidal wave is limited, thus in Eq. (2.57), both constants D_1, D_2 should be different from zero. Therefore we have:

$$\begin{aligned} \frac{3\hat{\beta}}{\hat{\alpha}}\left(\frac{\partial\zeta}{\partial x}\right)^2 &= -\zeta^3 + \frac{3}{\hat{\alpha}}\left(\frac{U}{C_0} - 1\right)\zeta^2 - \left(\frac{6}{C_0\hat{\alpha}}\right)D_1 - \zeta + \\ &+ \left(\frac{6}{C_0\hat{\alpha}}\right)D_2 = (\zeta_3 - \zeta)(\zeta - \zeta_2)(\zeta - \zeta_1) = P_3 \end{aligned} \tag{2.70}$$

in which $\zeta_1 < \zeta_2 < \zeta_3$ are zeros of the third-order polynomial P_3. In particular, ζ_3 and ζ_2 correspond to the heights of the crest and trough, respectively, and the difference $\zeta_3 - \zeta_2 = H$ is equal the total wave height. Following Mei et al. (2005) we find:

$$\zeta_1 = -\frac{H}{m}\frac{\mathbf{E}}{\mathbf{K}} \tag{2.71}$$

$$\zeta_2 = \frac{H}{m}\left(1 - \frac{\mathbf{E}}{\mathbf{K}} - m\right) \tag{2.72}$$

$$\zeta_3 = \frac{H}{m}\left(1 - \frac{\mathbf{E}}{\mathbf{K}}\right) \tag{2.73}$$

where **K** and **E** are the complete elliptic integrals of the first and second kind, respectively (Abramowitz and Stegun 1975):

$$\mathbf{K}(m) = \int_0^{\pi/2} \frac{du}{\sqrt{1 - m\sin^2 u}} \tag{2.74}$$

and

$$\mathbf{E}(m) = \int_0^{\pi/2} \sqrt{1 - m\sin^2 u}\, du \tag{2.75}$$

in which m is the parameter of the elliptic functions and integrals, such that $m = (\zeta_3 - \zeta_2)/(\zeta_3 - \zeta_1), 0 \leq m \leq 1$. In practice, the parameter m is not known "a priori" but must be obtained as a function of wave height H, wavelength L (or wave period T) and water depth h.

The solution of the KdV equation now becomes:

$$\begin{aligned}\zeta(x,t) &= \zeta_2 + Hcn^2\left[\left(\frac{\hat{\alpha}}{12\hat{\beta}}\right)^{1/2}(\zeta_3-\zeta_1)^{1/2}(x-Ut)\right] = \\ &= \frac{H}{m}\left(1-\frac{\mathbf{E}}{\mathbf{K}}-m\right) + Hcn^2\left[\left(\frac{\hat{\alpha}}{12\hat{\beta}}\right)^{1/2}\left(\frac{H}{m}\right)^{1/2}(x-Ut)\right] = \\ &= \frac{H}{m}\left(1-\frac{\mathbf{E}}{\mathbf{K}}-m\right) + Hcn^2\left[\frac{2\mathbf{K}(m)}{L}(x-Ut)\right]\end{aligned} \tag{2.76}$$

in which $cn(x,t;m)$ is the cosine-elliptic function which is, by definition, periodic with x with the period $4\mathbf{K}$ (Abramowitz and Stegun 1975).

Since $cn^2(x,t;m)$ must have the period $2\mathbf{K}(m)$, the wavelength L is:

$$L = \left(\frac{48\hat{\beta}}{\hat{\alpha}}\right)^{1/2}\left(\frac{m}{H}\right)^{1/2}\mathbf{K}(m) \tag{2.77}$$

or

$$L^2 = \frac{16}{3}\frac{h_r^3}{H}m\mathbf{K}^2(m) \tag{2.78}$$

and

$$\left(\frac{H}{h_r}\right)\left(\frac{L}{h_r}\right)^2 = \frac{16}{3}m\mathbf{K}^2(m) \tag{2.79}$$

which are transcendental relationships for unknown parameter m. For cnoidal wave speed we have:

$$U = C_0\left\{1+\frac{\hat{\alpha}H}{3m}\left[2-3\frac{\mathbf{E}}{\mathbf{K}}-m\right]\right\} \tag{2.80}$$

or

$$U = C_0\left\{1+\frac{1}{2}\frac{h_1-h_2}{h_1h_2}\left(\frac{H}{m}\right)\left[2-3\frac{\mathbf{E}}{\mathbf{K}}-m\right]\right\} \tag{2.81}$$

When $m \to 1$, we have $\mathbf{K}(1) = \infty$, $\mathbf{E}(1) = 1$. Thus $\zeta_2 \to \zeta_1$, $L \to \infty$ and $cn^2(x) \to \cosh^{-2}(x)$. Taking into account that the ratio $\mathbf{K}/L$ approaches a finite limit (see Eq. 2.77), from Eq. (2.76) we get the solitary wave profile:

$$\zeta(x,t) = H \cosh^{-2}\left\{\sqrt{\frac{3H}{4h_r^3}}(x-Ut)\right\} \tag{2.82}$$

which is equivalent to Eq. (2.65) when we use $H = A$.

From Eq. (2.80) for $m \rightarrow 1$ it follows that the wave speed U simplifies to Eq. (2.61) as for the solitary waves should be.

In the second limiting case when $m \rightarrow 0$, we have $\zeta_3 - \zeta_2 = H \rightarrow 0$, $\mathbf{K} = \mathbf{E} = \pi/2$, $U \rightarrow C_0$, $cn(u; m) \rightarrow \cos(u)$ and $\zeta_2 = -H/2$. Therefore:

$$\zeta = \frac{H}{2} \cos\left[\frac{2\pi}{L}(x - C_0 t)\right] = A\cos(kx - \omega t) \tag{2.83}$$

In Fig. 2.4 the surface elevations of the solitary wave (a) and cnoidal internal wave (b) are illustrated for the following wave parameters: wave height $H = 5$ m, wavelength $L = 500$ m, water depth $h = 80$ m, depth of interface $h_1 = 70$ m, and densities of the upper and lower layer $\rho_1 = 1004\,\mathrm{kg\,m^{-3}}$, $\rho_2 = 1012\,\mathrm{kg\,m^{-3}}$, respectively. The resulting wave parameters become: the reduced depth $h_r = 20.14$ m (see Eq. 2.64), linear wave speed $C_0 = 0.81\,\mathrm{m\,s^{-1}}$, wave speed $U = 1.41\,\mathrm{m\,s^{-1}}$ (for cnoidal wave), depth of wave trough $\zeta_2 = -0.94$ m and wave period $T = 354$ s.

Till now we have not discussed the breaking of the long waves. However, Grue et al. (2000) developed a fully nonlinear two-layer model where a thin layer of linearly stratified fluid has constant Väisälä-Brunt frequency and a thick layer has zero Väisälä-Brunt frequency. Particular attention was paid to the role of the breaking of waves. For large waves, the breaking occurs in a region in the centre of the waves, and breaking limits the fluid velocity. The experimental and theoretical velocities exhibit a good agreement up to breaking. The maximal amplitude in the experiments was about 1.25 times the depth of the linearly stratified layer, and the maximal propagation speed was approximately 1.78 times the linear long wave speed.

Two-layer model of internal waves described in this chapter should be considered only as an approximation to the real vertical distribution density. In general, the observations in natural conditions showed that the vertical distribution of water density is an arbitrary function of vertical coordinate z. Sometimes, for practical purposes, the following characteristic parameters of the two-layer stratification are used:

$$\chi = \frac{h_p}{h}, \quad \sigma = \frac{\Delta}{\rho_0} \tag{2.84}$$

in which h_p is the scale of thickness of the pycnocline, $\Delta = \rho_2 - \rho_1$, and ρ_0 is the reference density. The parameter χ is known as the Boussinesq parameter. If $\chi \ll 1$, the system is nearly two-layered. Usually in the natural conditions, the parameter χ is much bigger, maybe except the laboratory experiments, and the system should be considered as multi-layered, as is shown in the next chapter.

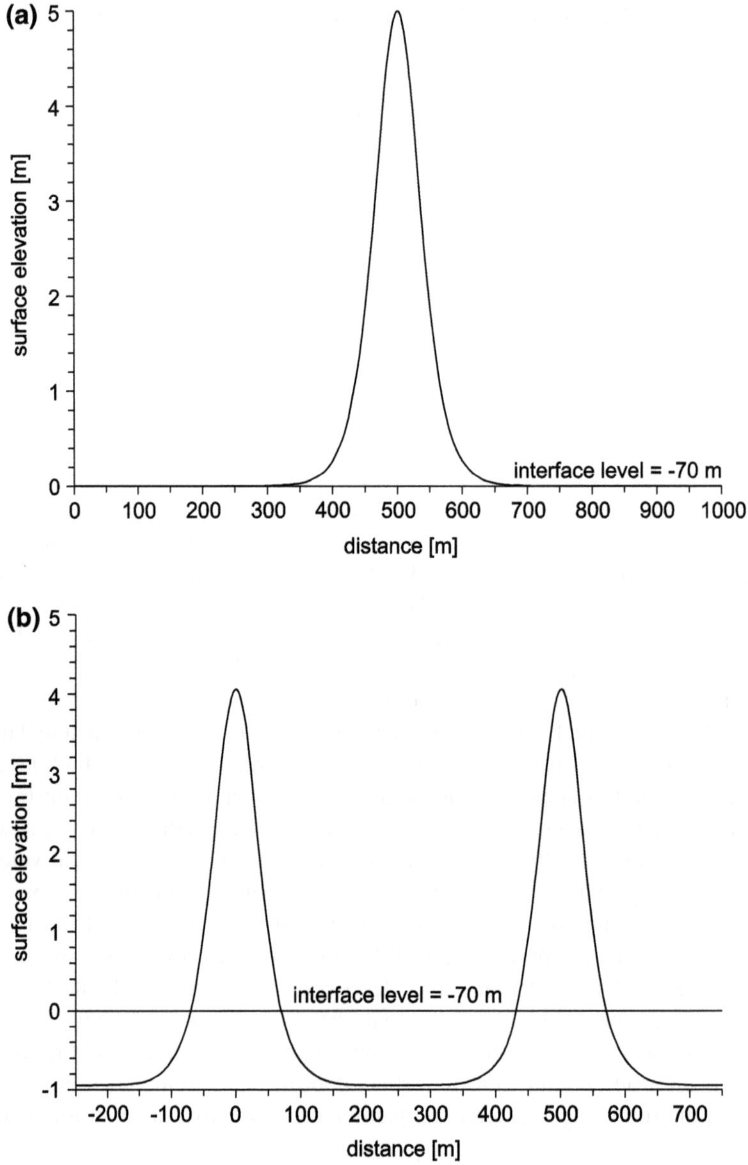

Fig. 2.4 Surface elevations of solitary (**a**) and cnoidal waves (**b**) of the same wave height

References

Abramowitz M, Stegun IA (1975) Handbook of mathematical functions. Dover Publications, New York, 1045 p

Grimshaw R, Pelinovsky E, Talipova T, Kurkin A (2004) Simulation of the transformation of internal solitary waves on oceanic shelves. J Phys Oceanogr 34(12):2774–2791

Grimshaw R, Pelinovsky E, Talipova T (2007) Modelling internal solitary waves in the coastal ocean. Surv Geophys 28:273–298

Grimshaw R, Guo C, Helfrich K, Vlasenko V (2014) Combined effect of rotation and topography on shoaling oceanic internal solitary waves. J Phys Oceanogr 44(4):1116–1132

Grue J, Jensen A, Rusas POV, Sveen JK (2000) Breaking and broadening of internal solitary waves. J Fluid Mech 413:181–217

Helfrich KR, Melville WK (1986) On long nonlinear internal waves over slope-shelf topography. J Fluid Mech 167:285–308

Helfrich KR, Melville WK (2006) Long nonlinear waves. Annu Rev Fluid Mech 38:395–425

Hunt JN (1961) Interfacial waves of finite amplitude. La Houille Blanche 4:515–531

Korteweg DJ, de Vries G (1895) On the change of form of long waves advancing in a rectangular canal, and on a new type of long stationary waves. Phil Mag and J Science 39:422–443

Lien RC, Henyey F, Ma B (2014) Large-amplitude internal solitary waves observed in the Northern South China Sea: properties and energetics. J Phys Oceanogr 44(4):1095–1115

Liu C-M (2006) Second-order random internal and surface waves in a two-fluid system. Geophys Res Lett 33:L06610

Massel SR (1989) Hydrodynamics of coastal zones. Elsevier, Amsterdam, 336 p

Massel SR (1999) Fluid mechanics for marine ecologists. Springer, Berlin, 566 p

Mei CC, Stassine M, Yue DK-P (2005) Theory and applications of ocean surface waves. World Scientific Publishing, Singapore, 1071 p

Miles JW (1981) The Korteweg-de Vries equation: a historical essay. J Fluid Mech 106:131–147

Pelinovsky E, Talipova T, Stepanyants Yu (1994) Modelling of nonlinear wave propagation in the horizontally inhomogeneous ocean. Fizika Atm i Okeana 30:79–85 (in Russian)

Pelinovsky E, Poluhina O, Slunyaev A (2007) Solitary waves in fluids. In: Grimshaw R (ed) Internal solitary waves. WIT Press, Southampton, pp 85–110

Piechura J, Beszczyńska-Möller A (2004) Inflow waters in the deep regions of the southern Baltic Sea-transport and transformations. Oceanologia 46(1):113–141

Pinet PR (1992) Oceanography: an introduction to the planet oceans. West Publishing Company, New York, 571 p

Sharma JN, Dean RG (1979) Development and evaluation of a procedure for simulating a random directional second order sea surface and associated wave forces, vol 20, Ocean Engineering Report University of Delaware, Newark

Song J-B (2004) Second-order random wave solutions for internal waves in a two-layer fluid. Geophys Res Lett 31:L15302

Sutherland B (2010) Internal gravity waves. Cambridge University Press, Cambridge, 377 p

Thorpe SA (1968) On the shape of progressive internal waves. Philos T Roy Soc A 263:563–614

Whitham GB (1974) Linear and nonlinear waves. Wiley, New York, 636 p

Zheng Q, Klemas V, Yan XH, Pan J (2001) Nonlinear evolution of ocean internal solitons propagating along an inhomogeneous thermocline. J Geophys Res 106(C7):14083–14094

References

[illegible]

Chapter 3
Waves in a Continuously Stratified Sea of Constant Water Depth

3.1 Introduction

Ocean water is continuously stratified, i.e. the water density varies continuously with the depth. Internal waves still occur, but in contrast to the two-layer model, internal waves can also propagate in non-horizontal direction. Thus, wavenumber should be treated as the vector $\mathbf{k} = \mathbf{k}(k_x, k_y, k_z)$ in three dimensional space. This fact is of basic importance for the dynamics of ocean water and biological life in the ocean, as internal waves two-layer model, internal waves can also propagate in non-horizontal direction. propagating in the vertical direction transport energy and nutrients from the surface to the bottom of the ocean and vice versa. The internal waves dynamics depend very strongly on the vertical structure of the water column expressed in terms of density and Väisälä-Brunt frequency.

In the following Sections the various types of density profiles are examined under the assumption of constant water depth. In Sect. 3.2, a sample case of exponential density profile is considered. As the Väisälä-Brunt frequency is constant, a closed analytical solution is possible. For shallow seas such as the Baltic Sea, the vertical density structure exhibits the three-layer profile when surface and bottom uniform layers are separated by a non-uniform layer. If the density profile in the non-uniform layer is approximated by the exponential function, the boundary value problem for the three-layer density profile was solved analytically in Sect. 3.3. Some other density profiles are discussed in Sect. 3.4.

For a more complex density profile, an analytical solution is not possible. In Sect. 3.5, the variational method for arbitrary density profile is discussed and its applicability is demonstrated. A specific case of two-layer system of lower pycnocline and upper seasonal pycnocline, and exchange energy between them is described shortly in Sect. 3.6. For long, weakly nonlinear waves, a different approach is required. Similarly to the two-layer case, internal wave motion can be described by the KdV equation. The solitary and cnoidal type solutions of this equation are developed and illustrated by some examples in Sect. 3.7. Finally, the considerations of the energy and mixing due to internal waves are given shortly in Sects. 3.8 and 3.9.

S.R. Massel, *Internal Gravity Waves in the Shallow Seas*,
GeoPlanet: Earth and Planetary Sciences, DOI 10.1007/978-3-319-18908-6_3

3.2 Exponential Density Profile

The simplest case of the density profile is for the Väisälä-Brunt frequency being constant in the whole water column, i.e. $N(z) = N_0$, which makes governing equations for internal waves analytically tractable. The constant Väisälä-Brunt frequency corresponds to the density profile in the form:

$$\overline{\rho}(z) = \overline{\rho}_0 \exp\left(-\frac{N_0^2}{g}z\right) \tag{3.1}$$

Substituting Eq. (3.1) into Eq. (1.7) we obtain the Väisälä-Brunt frequency as follows:

$$N^2(z) = -\frac{g}{\overline{\rho}}\frac{d\overline{\rho}}{dz} = N_0^2 = \text{const} \tag{3.2}$$

Allowing waves of frequency ω to propagate in the constant water depth and in the vertical plane (x, z), i.e. when $k_h^2 = k_x^2$, Eqs. (1.46) and (1.47) give:

$$\frac{d^2W}{dz^2} + \frac{N_0^2 - \omega^2}{\omega^2 - f^2}k_x^2\, W = 0 \tag{3.3}$$

and

$$\frac{d^2F}{dx^2} + k_x^2\, F = 0 \tag{3.4}$$

The general solution of equation (3.3) is:

$$W(z) = C\cos(k_z z) + B\sin(k_z z) \tag{3.5}$$

where:

$$k_z^2 = \frac{N_0^2 - \omega^2}{\omega^2 - f^2}k_x^2 \tag{3.6}$$

or

$$\omega^2 = \frac{k_x^2(N_0^2 - f^2)}{k_x^2 + k_z^2} + f^2 = \frac{N_0^2 + f^2\left(\dfrac{k_z}{k_x}\right)^2}{1 + \left(\dfrac{k_z}{k_x}\right)^2} \tag{3.7}$$

Note that the dispersion relation is independent of the boundary conditions.

If $f^2 \ll \omega^2$ and waves are of short period, the Earth's rotation may be neglected. Thus Eq. (3.7) becomes:

$$\left(\frac{\omega}{N_0}\right)^2 = \frac{k_x^2}{k_x^2 + k_z^2} \tag{3.8}$$

From Eq. (3.6) it follows that the vertical wavenumber k_z is given by:

$$k_z = \pm\sqrt{\frac{N_0^2 - \omega^2}{\omega^2 - f^2}}\, k_x \tag{3.9}$$

If waves propagate upwards into a region of increasing frequency N_0 in the thermocline, wavenumber k_z increases as $\sqrt{N^2 - \omega^2}$ and the wavelength $\lambda = 2\pi/(k_x^2 + k_z^2)^{1/2}$ is reduced.

The magnitude of vertical energy flux $F_z = EC_{gz}$ becomes (Thorpe 2005):

$$|F_z| = \frac{\rho_0 a^2 \omega (N_0^2 - \omega^2)^{1/2} (\omega^2 - f^2)^{1/2}}{2k_x} \tag{3.10}$$

in which a is the amplitude of internal wave and ρ_0 is a reference density. In absence of mean shear and negligible dissipation, the wave amplitude, a, varies roughly as $N_0^{-1/2}$, and attenuates as upward propagating waves encounter the thermocline (Thorpe 2005). However, when frequency N_0 decreases to values less than ω, wavenumber k_z tends to infinity and wave are reflected. Therefore, wave energy may be effectively trapped within regions where N is large.

When inertial frequency $f = 0$, from Eq. (3.10) we have:

$$|F_z| = \frac{\rho_0 a^2 \omega^2 (N^2 - \omega^2)^{1/2}}{2k} \tag{3.11}$$

The maximum of vertical energy flux appears when $\omega/N = (2/3)^{1/2}$. It means from Eq. (1.54) that angle $\theta = \arcsin(2/3)^{1/2} \approx 54.7°$.

The internal waves produce only very small vertical displacements of the free surface, therefore the free surface condition at the sea surface is usually taken as:

$$w = 0 \quad \text{at} \quad z = 0 \tag{3.12}$$

At the impermeable sea bottom the normal velocity should vanish, so in the special case of the horizontal bottom we have:

$$w = 0 \quad \text{at} \quad z = -h \tag{3.13}$$

The surface boundary condition (3.12), induces that $C = 0$ in Eq. (3.5). On the other hand, at the sea bottom should be $\sin[k_z(-h)] = 0$. Thus:

$$k_z^{(n)} h = n\pi \quad \text{and} \quad k_z^{(n)} = \frac{n\pi}{h} \tag{3.14}$$

The symbol n denotes the number of mode. Functions $W(z)$ and $F(x)$ for particular mode now become:

$$W^{(n)}(z) = B^{(n)} \sin(k_z^{(n)} z) = B^{(n)} \sin\left(\frac{n\pi z}{h}\right) \tag{3.15}$$

and

$$F^{(n)}(x) = \exp(ik_x^{(n)} x) \tag{3.16}$$

where:

$$k_x^{(n)} = \frac{n\pi}{h}\sqrt{\frac{\omega^2 - f^2}{N_0^2 - \omega^2}} \tag{3.17}$$

From Eq. (3.17) we find the horizontal component of phase velocity $C_{0x}^{(n)}$ as follows:

$$C_{0x}^{(n)} = \sqrt{\frac{N_0^2 - \omega^2}{\omega^2 - f^2}}\frac{\omega h}{n\pi} \tag{3.18}$$

while the vertical component of phase velocity $C_{0z}^{(n)}$ is:

$$C_{0z}^{(n)} = \frac{\omega h}{n\pi} \tag{3.19}$$

Therefore, the aspect ratio becomes:

$$\alpha = \frac{k_x^{(n)}}{k_z^{(n)}} = \frac{C_{0z}^{(n)}}{C_{0x}^{(n)}} = \sqrt{\frac{\omega^2 - f^2}{N_0^2 - \omega^2}} \tag{3.20}$$

For short internal waves, Eqs. (3.18)–(3.20) simplify as follows:

$$C_{0x}^{(n)} = \frac{\sqrt{N_0^2 - \omega^2}h}{n\pi}, \quad C_{0z}^{(n)} = \frac{\omega h}{n\pi}, \quad \alpha = \frac{\omega}{\sqrt{N_0^2 - \omega^2}} \tag{3.21}$$

and for long waves we have:

$$C_{0x}^{(n)} = \frac{N_0}{\sqrt{\omega^2 - f^2}}\frac{\omega h}{n\pi}, \quad C_{0z}^{(n)} = \frac{\omega h}{n\pi}, \quad \alpha = \frac{\sqrt{\omega^2 - f^2}}{N_0} \tag{3.22}$$

It should be noted, when the Earth rotation is neglected ($f = 0$), the long waves are non-dispersive and both phase velocity components are independent of frequency ω.

The final solution for nth mode of the vertical velocity for a given wave frequency ω now is:

$$w^{(n)}(x, z, t) = \Re B^{(n)} \sin\left(\frac{n\pi z}{h}\right) \exp[i(k_x^{(n)} x - \omega t)] \tag{3.23}$$

in which the still unknown velocity amplitude $B^{(n)}$ depends on forcing of the internal waves due to external processes and $\Re$ denotes the real part of a given expression.

Let us now present vertical and horizontal velocities, $u^{(n)}(x, z, t)$ and $w^{(n)}(x, z, t)$ in more general forms:

$$u^{(n)}(x, z, t) = \Re U^{(n)}(z) \exp[i(k_x^{(n)} x)] \exp(-i\omega t) \tag{3.24}$$

and

$$w^{(n)}(x, z, t) = \Re W^{(n)}(z) \exp[i(k_x^{(n)} x)] \exp(-i\omega t) \tag{3.25}$$

The horizontal velocity can be obtained from the equation of continuity (1.35) when no y-dependence is assumed, i.e. $\partial u/\partial x + \partial w/\partial z = 0$. Thus we have:

$$u^{(n)}(x, z, t) = \Re \frac{1}{k_x^{(n)}} \frac{dW^{(n)}}{dz} \exp\left[i\left(k_x^{(n)} x - \omega t + \frac{\pi}{2}\right)\right] \tag{3.26}$$

Thus the horizontal velocity amplitude is proportional to the first derivative of vertical velocity. Substituting Eq. (3.15) into Eq. (3.26), for a given frequency we get ω:

$$u^{(n)}(x, z, t) = \Re B^{(n)} \frac{n\pi}{k_x^{(n)} h} \cos\left(\frac{n\pi z}{h}\right) \exp\left[i\left(k_x^{(n)} x - \omega t + \frac{\pi}{2}\right)\right] \tag{3.27}$$

or

$$u^{(n)}(x, z, t) = \Re U^{(n)}(z) \exp\left[i\left(k_(n)_x x - \omega t + \frac{\pi}{2}\right)\right] \tag{3.28}$$

where:

$$U^{(n)}(z) = B^{(n)} \left(\frac{n\pi}{k_x^{(n)} h}\right) \cos\left(\frac{n\pi z}{h}\right) = B^{(n)} \left(\frac{k_z^{(n)}}{k_x^{(n)}}\right) \cos(k_z^{(n)} z) \tag{3.29}$$

Equations (3.23) and (3.27) show that there is only a series of allowable modes and frequencies of internal waves. For illustration of the modal structure of the vertical and horizontal velocities, the first three modes ($n = 1, 2, 3$) are shown in Fig. 3.1. Note that the mode number, n, equals the number of zero crossings for the velocity, u, or the number of extrema (maxima and minima) for the velocity w.

Figure 3.1a suggests that the amplitude of the first mode of the velocity component, $w^{(1)}$, is positive for all water levels (constant $B^{(1)}$ is assumed to be negative).

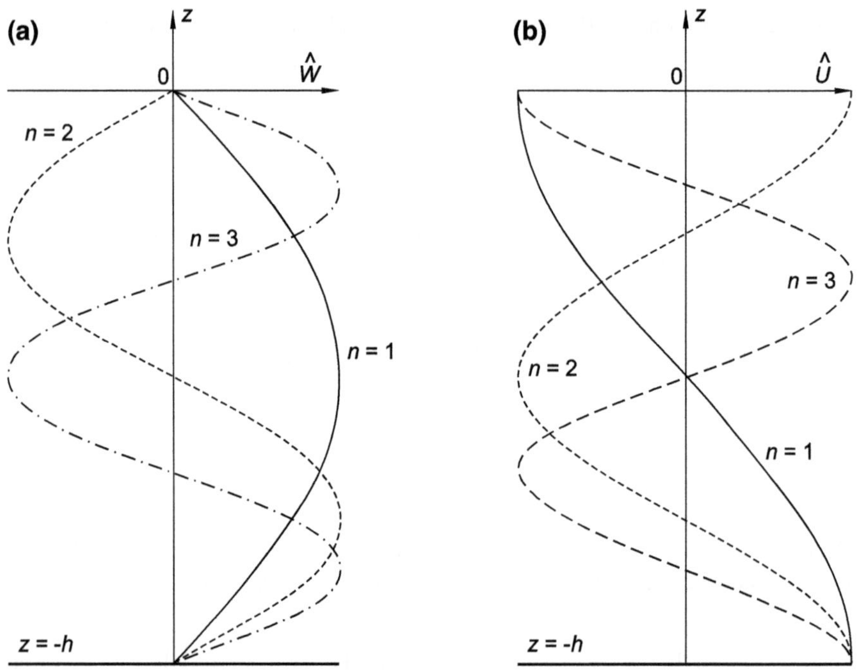

Fig. 3.1 Vertical profiles of velocity amplitudes for the first three modes of the internal waves, when N is constant: **a** vertical component, **b** horizontal component

Therefore, when the phase $k_x^{(1)}x - \omega t < \pi$, the vertical velocity $w^{(1)}$ is directed upwards. For the second ($n = 2$) and third mode ($n = 3$), the velocity $w^{(n)}$ is expected to be positive only for $-h/2 < z < 0$, and for $-2h/3 < z < 0$, respectively. Similar observations can be made for the other wave phases and the $u^{(n)}$ component of velocity.

Because of the linear problem formulation, arbitrary sums of modes of velocities $w^{(n)}(x, z, t)$ and $u^{(n)}(x, z, t)$ are also solution of the Eq. (3.3), i.e.:

$$w(x, z, t) = \sum_n w^{(n)}(x, z, t) = \Re \sum_n B^{(n)} \sin\left(\frac{n\pi z}{h}\right) \exp[i(k_x^{(n)}x - \omega t)] \tag{3.30}$$

and

$$u(x, z, t) = \sum_n u^{(n)}(x, z, t) = \Re \sum_n B^{(n)} \frac{n\pi}{k_x^{(n)} h} \cos\left(\frac{n\pi z}{h}\right) \exp\left[i\left(k_x^{(n)}x - \omega t + \frac{\pi}{2}\right)\right] \tag{3.31}$$

By definition $w^{(n)} = \partial\zeta^{(n)}/\partial t$. Therefore, the vertical displacement of isopycnals $\zeta^{(n)}(x, z, t)$ becomes:

$$\zeta(x,z,t) = \sum_n \zeta^{(n)}(x,z,t) = \Re \sum_n \frac{B^{(n)}}{\omega} \sin\left(\frac{n\pi z}{h}\right) \exp\left[i\left(k_x^{(n)}x - \omega t + \frac{\pi}{2}\right)\right] \tag{3.32}$$

For periodic internal waves of frequency ω, the dynamic part of density $\rho(x, x, t)$ is also function of time, i.e.: $\rho \sim \exp(-i\omega t)$. From Eq. (1.22) we have:

$$\rho(x,z,t) = \frac{-iw(x,z,t)}{\omega}\frac{d\bar{\rho}}{dz} \tag{3.33}$$

Therefore, the total density is:

$$\rho^*(x,z,t) = \overline{\rho}(z) - \frac{iw^{(n)}(x,z,t)}{\omega}\frac{d\overline{\rho}(z)}{dz} \tag{3.34}$$

If density $\overline{\rho}(z)$ is given by Eq. (3.1), we have:

$$\rho^*(x,z,t) = \Re\overline{\rho}_0 \exp\left(-\frac{N_0^2 z}{g}\right)\left[1 + i\frac{N_0^2}{g\omega} w^{(n)}(x,z,t)\right] \tag{3.35}$$

After substituting the vertical velocity $w(x, z, t)$ from Eq. (3.23) into Eq. (3.35) we write the total density field due to nth mode of the internal wave in the form:

$$\rho_n^*(x,z,t) = \overline{\rho}_0 \exp\left(-\frac{N_0^2 z}{g}\right)\left[1 - \frac{B^{(n)}}{\omega}\frac{N_0^2}{g} \sin\left(\frac{n\pi z}{h}\right) \sin\left(k_x^{(n)}x - \omega t\right)\right] \tag{3.36}$$

We illustrate an application of the above formulae for a station in the Southern Baltic. The observation point is located at the latitude $\varphi = 55°$ and water depth $h = 100$ m. The corresponding inertial frequency $f = 1.19 \times 10^{-4}$ rad s^{-1} (T_{in} = 14.6 h). The density vertical profile is described approximately by Eq. (3.1), in which $\overline{\rho}(z = 0) = 1004$ kg m^{-3} and $\overline{\rho}(z = -100) = 1012$ kg m^{-3}. Thus the Väisälä-Brunt frequency becomes $N_0 = 0.027902$ rad s^{-1}(T $\approx$ 225 s). Additionally, we assume that frequency of the internal waves is $\omega = 6.98 \times 10^{-3}$ rad s^{-1} ($T = 0.25$ h).

Figure 3.2 illustrates the potential density σ_t for the first wave mode ($n = 1$) as a function of distance. The values of the corresponding wavenumbers are: $k_z^{(1)} = 0.0314$ m^{-1} and $k_x^{(1)} = 8.12 \times 10^{-3}$ m^{-1}. The value of constant $B^{(1)} = 0.2$ m s^{-1} was used in calculations. The shape of the isopycnals changes with water depth. Close to the sea surface the crests of the internal waves are more flattered.

In Fig. 3.3 the density σ_t is given for the superposition of first three modes when $B^{(1)} = B^{(2)} = B^{(3)} = 0.2$ m s^{-1}. All other initial data are the same as in Fig. 3.2. The inclusion of the higher modes results in the sharpening of the isopycnal slopes.

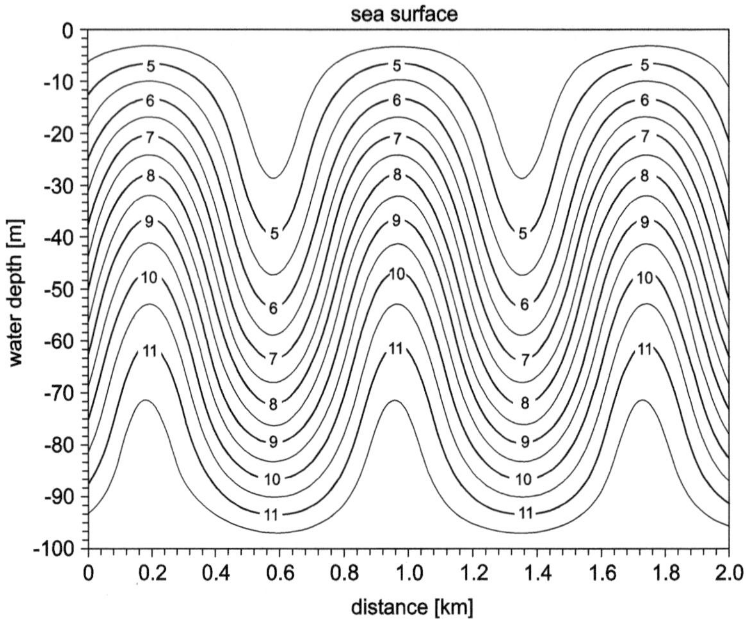

Fig. 3.2 Potential density of the first mode ($n = 1$) as a function of distance

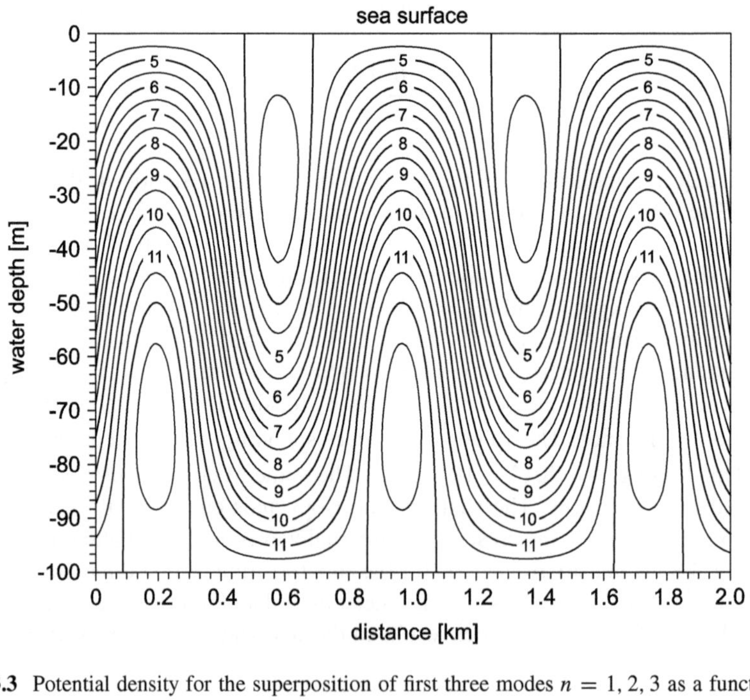

Fig. 3.3 Potential density for the superposition of first three modes $n = 1, 2, 3$ as a function of distance

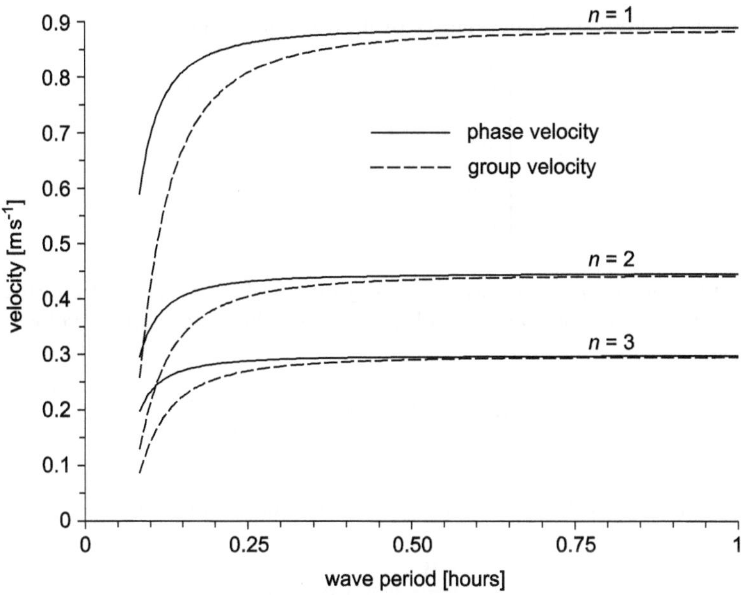

Fig. 3.4 Horizontal components of phase and group velocities as a function of wave period

Moreover, Fig. 3.4 illustrates the horizontal components of the phase and group velocities of the internal waves of periods 5 min $< T <$ 1 h. For this period range and water depth $h = 100$ m, the internal waves can be considered as short waves, maybe except the periods longer than about 0.75 h, when the waves should be considered as long ones.

3.3 Three-Layer Density Profile

The vertical density distribution with the upper and lower homogeneous layers and non-uniform layer between them is frequently observed in nature (Krauss 1966; Golenko et al. 1999; Kurkina et al. 2011). Therefore the three-layer model of the density profile is used to represent the internal waves in some sea basins, such as the Baltic Sea. Thus, we consider the following density distribution in water column (z-axis is directed upwards):

$$\overline{\rho}(z) = \begin{cases} \rho_1 & \text{for } -h_1 \le z \le 0 \\ \rho_1 \exp\left[-\dfrac{N_0^2}{g}(z+h_1)\right] & \text{for } -h_2 \le z \le -h_1 \\ \rho_2 & \text{for } -h \le z \le -h_2 \end{cases} \tag{3.37}$$

Therefore, the corresponding Väisälä-Brunt frequency N_0 is:

$$N_0^2 = \begin{cases} 0 & \text{for } -h_1 \le z \le 0 \\ \dfrac{g \ln\left(\dfrac{\rho_1}{\rho_2}\right)}{h_1 - h_2} & \text{for } -h_2 \le z \le -h_1 \\ 0 & \text{for } -h \le z \le -h_2 \end{cases} \tag{3.38}$$

Assuming that the internal wave is propagating in the (x, z) plane and the Earth rotation is taken into account, from Eq. (1.47) we obtain the vertical velocity W in the form:

$$\frac{d^2W}{dz^2} + \frac{N_0^2 - \omega^2}{\omega^2 - f^2} k_x^2 W = 0 \tag{3.39}$$

Following Krauss (1966), we rewrite the above equation in the form:

$$\frac{d^2W}{dz^2} + \beta^2 \nu W = 0 \tag{3.40}$$

in which:

$$\nu = \frac{k_x^2}{\omega^2 - f^2}, \quad \beta = \sqrt{N_0^2 - \omega^2} \tag{3.41}$$

In the particular layers, we adopt the velocity amplitudes as follows:

$$\left.\begin{array}{ll} W_1(z) = A \sinh(\sqrt{\nu}\omega z), & -h_1 \le z \le 0 \\ W_2(z) = B \sin(\beta\sqrt{\nu}z) + C\cos(\beta\sqrt{\nu}z), & -h_2 \le z \le -h_1 \\ W_3(z) = D \exp(\sqrt{\nu}\omega z) + E\exp(-\sqrt{\nu}\omega z), & -h \le z \le -h_2 \end{array}\right\} \tag{3.42}$$

Equation (3.42) should satisfy the following boundary conditions at the sublayers boundaries. Thus we have:

$$\left.\begin{array}{l} W_1(0) = W_3(-h) = 0 \\ W_1(-h_1) = W_2(-h_1), \quad \dfrac{dW_1(-h_1)}{dz} = \dfrac{dW_2(-h_1)}{dz} \\ W_2(-h_2) = W_3(-h_2), \quad \dfrac{dW_2(-h_2)}{dz} = \dfrac{dW_3(-h_2)}{dz} \end{array}\right\} \tag{3.43}$$

After substituting Eq. (3.42) into (3.43) we obtain the condition of the non-trivial solution in the form of following determinant Δ equation:

$$\Delta = \begin{vmatrix} a_{11} & a_{12} & a_{13} & a_{14} \\ a_{21} & a_{22} & a_{23} & a_{24} \\ a_{31} & a_{32} & a_{33} & a_{34} \\ a_{41} & a_{42} & a_{43} & a_{44} \end{vmatrix} = 0 \tag{3.44}$$

where:

$$\left.\begin{array}{ll} a_{11} = -\sinh(\sqrt{\overline{\nu}}\omega h_1), & a_{12} = \sin(\beta\sqrt{\overline{\nu}}h_1) \\ a_{13} = -\cos(\beta\sqrt{\overline{\nu}}h_1), & a_{14} = 0.0 \\ a_{21} = \omega\cosh(\sqrt{\overline{\nu}}\omega h_1), & a_{22} = -\beta\cos(\beta\sqrt{\overline{\nu}}h_1) \\ a_{23} = -\beta\sin(\beta\sqrt{\overline{\nu}}h_1), & a_{24} = 0.0 \\ a_{31} = 0.0, & a_{32} = -\sin(\beta\sqrt{\overline{\nu}}h_2) \\ a_{33} = \cos(\beta\sqrt{\overline{\nu}}h_2), & a_{34} = -\exp(-\sqrt{\overline{\nu}}\omega h_2)[1+\exp(-2\sqrt{\overline{\nu}}\omega h_1)] \\ a_{41} = 0.0, & a_{42} = \beta\cos(\beta\sqrt{\overline{\nu}}h_2) \\ a_{43} = \beta\sin(\beta\sqrt{\overline{\nu}}h_2), & a_{44} = -\omega\exp(-\sqrt{\overline{\nu}}\omega h_2)[1-\exp(-2\sqrt{\overline{\nu}}\omega h_1)] \end{array}\right\} \tag{3.45}$$

For a fixed ω, vertical velocity amplitudes $W^{(n)}(z)$ resulting from Eq. (3.40) can be determined for an infinite number of modes and corresponding horizontal wavenumbers $k^{(1)}, k^{(2)}, k^{(3)}, \ldots, k^{(n)}$. The internal wave characterised by nth wavenumber is called an nth-order internal wave, or an internal wave of the nth mode.

The equation $\Delta = 0$ allows to determine the dispersion relation and, in consequence, the phase velocity. From Eq. (3.44) it follows that phase velocity is independent on the amplitudes A, B, C, D and E. In Fig. 3.5 the phase velocities for the three-layer system are illustrated for the first three modes and internal wave period in the range $0.25 < T < 1.5$ h for typical conditions in the Gotland Deep in the Baltic Sea (Talipova et al. 1998). The characteristic water depths are adopted as follows: $h_1 = 60$ m, $h_2 = 200$ m and $h = 240$ m (Station 41). The densities in the upper and lower layers equal $\rho_1 = 1005.8\,\mathrm{kg\,m^{-3}}$ and $\rho_2 = 1009.0\,\mathrm{kg\,m^{-3}}$ respectively. The corresponding Väisälä-Brunt frequency N_0 from Eq. (3.38) is equal to $0.0149\,\mathrm{s^{-1}}$ ($T \approx 420$ s). It should be noted that in a given wave period range the phase velocity behaves as for a long, almost non-dispersive wave.

The vertical velocity $w(x, z, t)$ for each mode n has the form:

$$w(x, z, t) = \Re\, W(z) \exp[i(kx - \omega t)] \tag{3.46}$$

in which the amplitudes $W(z)$ for the particular layers are defined in Eq. (3.42). The unknown coefficients A, B, C, D and E should be found as a solution of the equation system (3.43). After substituting Eqs. (3.42) into (3.43) we obtain:

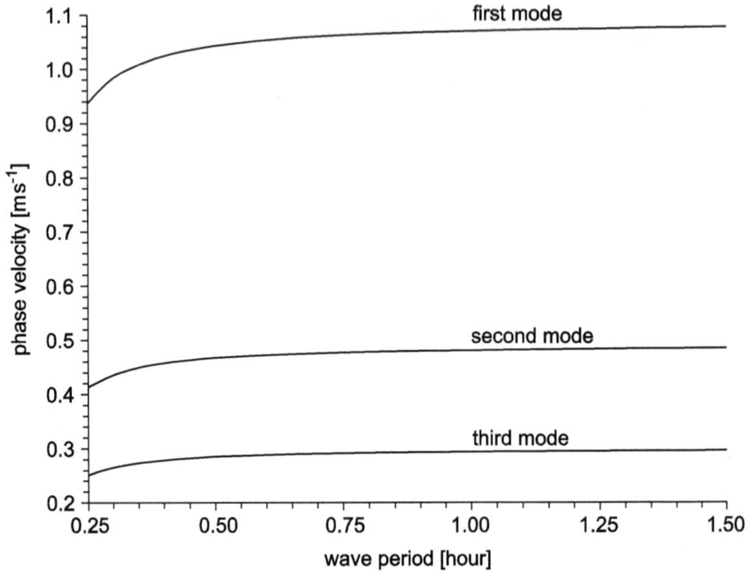

Fig. 3.5 Phase velocities as a function of wave period for three-layer model

$$\left.\begin{aligned} B &= \frac{a_{13}a_{21} - a_{11}a_{23}}{a_{12}a_{23} - a_{13}a_{22}} \cdot A \\ C &= \frac{a_{11}a_{22} - a_{12}a_{21}}{a_{12}a_{23} - a_{13}a_{22}} \cdot A \\ D &= \frac{-(a_{32}B + a_{33}C)}{a_{34}} \cdot A \\ E &= -\exp(-2\sqrt{\nu}\,\omega h) \cdot D \end{aligned}\right\} \tag{3.47}$$

The horizontal component velocity $u(x, z, t)$ can be obtained from the continuity equation, i.e.:

$$\frac{\partial u}{\partial x} + \frac{\partial w}{\partial z} = 0 \tag{3.48}$$

So, let us present the horizontal velocity $u(x, z, t)$ as:

$$u(x, z, t) = \Re\, G(z) \exp[i(kx - \omega t)] \tag{3.49}$$

and substitute it in Eq. (3.48):

$$u(x, z, t) = \Re \frac{i}{k} \frac{dW(z)}{dz} \exp[i(kx - \omega t)] \tag{3.50}$$

Now, using Eq. (3.42), we represent the amplitudes $G(z)$ for particular layers as follows:

$$\left.\begin{aligned} G_1(z) &= \Re A \frac{i\omega}{\sqrt{\omega^2 - f^2}} \cosh(\sqrt{\nu}\,\omega z) \ \text{ for } -h_1 \le z \le 0 \\ G_2(z) &= \Re\, i \sqrt{\frac{N^2 - \omega^2}{\omega^2 - f^2}} [B \cos(\beta\sqrt{\nu} z) - C \sin(\beta\sqrt{\nu} z)] \ \text{ for } -h_2 \le z \le -h_1 \\ G_3(z) &= \Re \frac{i\omega}{\sqrt{\omega^2 - f^2}} [D \exp(\sqrt{\nu}\,\omega z) - E \exp(-\sqrt{\nu}\omega z)] \ \text{ for } -h \le z \le -h_2 \end{aligned}\right\} \tag{3.51}$$

From the equations above it follows that wave induced velocities are determined with an correspondence to the constant velocity A which should be found from the external forcing.

Figure 3.6 shows the profiles of the vertical velocity amplitudes $W(z)$ for the first three modes of internal waves of period $T = 0.5$ h in the Gotland Deep (Station 16). From data given in Talipova et al. (1998), the following characteristic water depths have been selected: $h = 160$ m, $h_1 = 70$ m and $h_2 = 140$ m. In calculations the parameter A was assumed to be equal 0.01 m s^{-1}. The corresponding horizontal velocity amplitudes are given in Fig. 3.7.

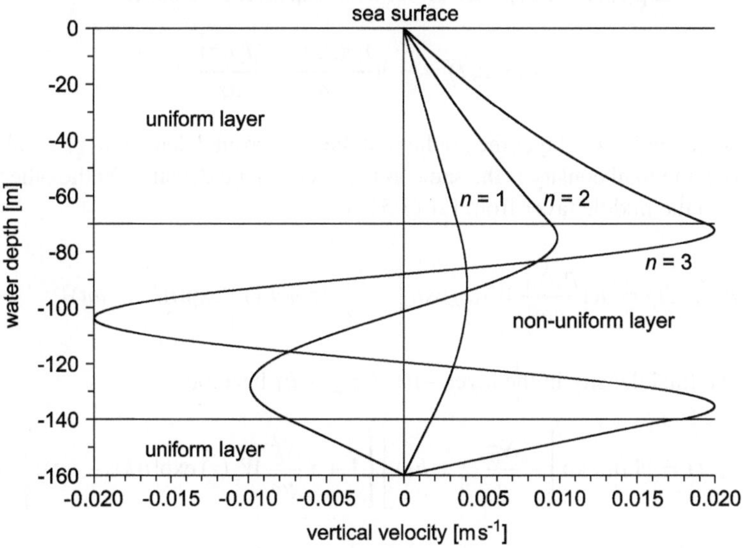

Fig. 3.6 Profiles of the vertical velocity amplitudes $W(z)$ for the three modes of waves of period $T = 0.5$ h

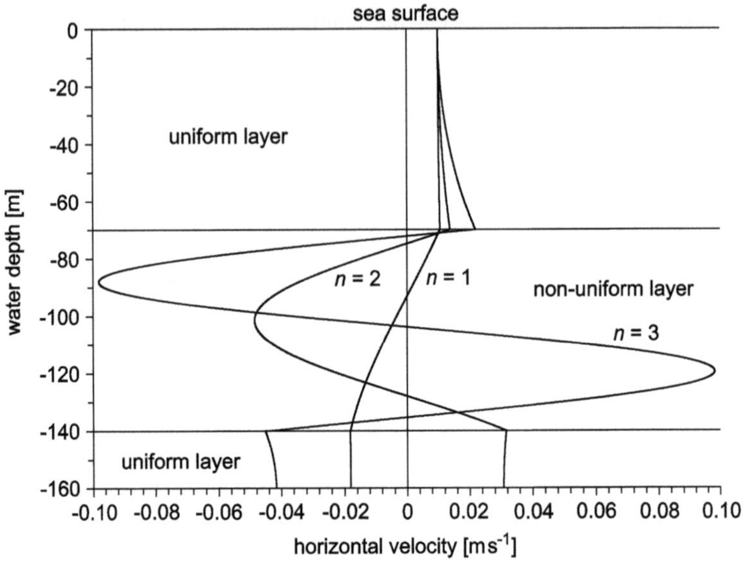

Fig. 3.7 Profiles of horizontal velocity amplitudes $G(z)$ for the three modes of waves of period $T = 0.5\,\mathrm{h}$

Final density field is a summation of water density without the waves $\bar{\rho}(z)$ and density variation due to internal wave motion $\rho(x, z, t)$. Let us assume that density $\rho(x, z, t) \sim \exp[i(kx - \omega t)]$. Therefore, from Eq. (1.22) we have:

$$\rho(x, z, t) = -\Re \frac{iw(x, z, t)}{\omega} \frac{d\bar{\rho}(z)}{dz} \tag{3.52}$$

In the upper and lower layer the gradient of the background density $\bar{\rho}(z)$ equals zero; therefore, the final density is the same as the background density. On the other hand, density in the middle layer from Eq. (3.52) is:

$$\rho(x, z, t) = \Re\, i \frac{\rho_1 N_0^2}{g\omega} W_2(z) \exp\left[-\frac{N_0^2}{g}(z + h_1)\right] \exp[i(kx - \omega t)] \tag{3.53}$$

Thus, the final density in the layer $-h_2 \le z \le -h_1$ becomes:

$$\rho_t(x, z, t) = \Re\, \rho_1 \exp\left[-\frac{N_0^2}{g}(z + h_1)\right]\left\{1 + i\frac{N_0^2}{g\omega} W_2(z) \exp[i(kx - \omega t)]\right\} \tag{3.54}$$

The water density in the non-uniform layer $h_2 = -140 \le z \le h_1 = -70$ was shown in Fig. 3.8 for first mode only. The internal wave heights reached about 2 m.

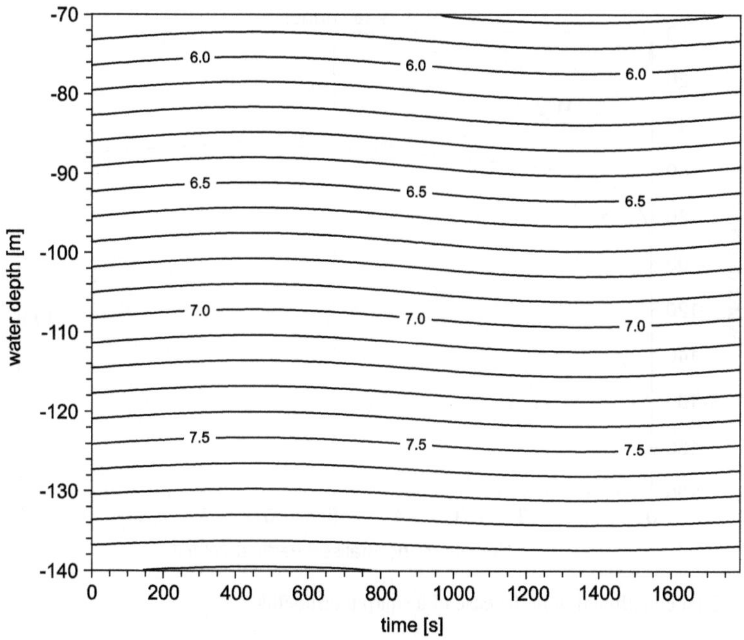

Fig. 3.8 Distribution of water density in the non-uniform water layer $-140\,\mathrm{m} \leq z \leq -70\,\mathrm{m}$

3.4 Other Specific Density Profiles

The assumption of the exponential density distribution (3.37) results in a constant Väisälä-Brunt frequency N and substantially simplifies the determination of the various internal wave parameters. Except the exponential density profile, there are other profiles for which an analytical solution is possible. Very instructive examples are listed in the monographs of Krauss (1966) and Roberts (1975).

In tropical or subtropical waters, when the upper layer of the ocean is well-mixed by wind, the fairly sharp thermocline separates nearly homogeneous water masses above and below. If we assume that waves are short enough and the effect of the Earth's rotation may be neglected ($f = 0$), solution of Eq. (3.40) takes the form:

$$W(z) = \begin{cases} C_1 \sinh(kz) & \text{for } -h_1 \leq z \leq 0 \\ C_2 \sinh[k(z+h)] & \text{for } -h \leq z \leq -h_1 \end{cases} \tag{3.55}$$

in which h is the water depth and h_1 is a central level of the non-homogeneous density profile.

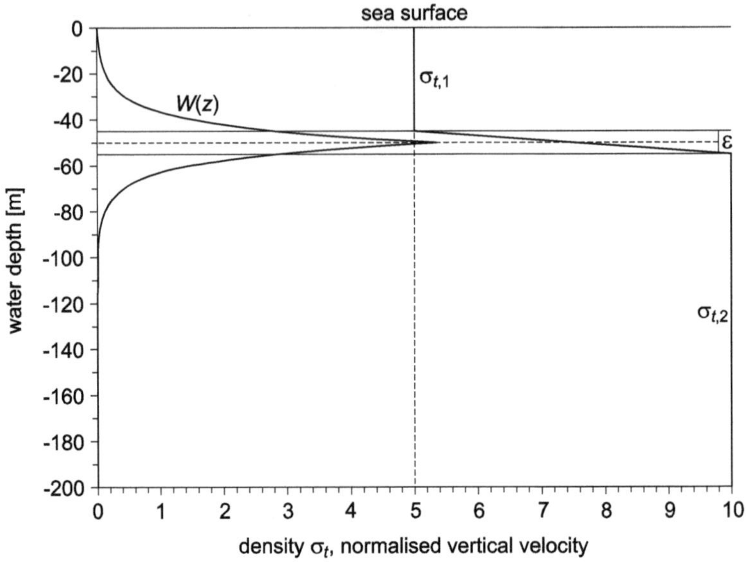

Fig. 3.9 First eigenfunction in the case of a sharp thermocline

The corresponding dispersion relation can be expressed as follows:

$$\omega^2 - gk\,\frac{\rho_2 - \rho_1}{\rho_1}\,\{k\varepsilon + \coth(kh) + \coth[(h - h_1)]\}^{-1} = 0 \tag{3.56}$$

where ε is the thermocline thickness.

In Fig. 3.9 an example of calculations for short internal wave and for sharp thermocline of the width $\varepsilon = 10\,\mathrm{m}$, located at water depth $h_1 = 50\,\mathrm{m}$ is given. The upper and lower densities are equal 1005.0 and $1010\,\mathrm{kg\,m^{-3}}$, respectively. However, care must be used when waves are long compared to the thermocline thickness.

Similar solution can be found for the non-homogeneous layer when the density profile takes the form:

$$\rho(z) = \rho(-h_1)\exp\left\{\frac{\rho_2 - \rho_1}{2g\rho(-h_1)}\tanh\left[-\frac{(z+h)}{2\varepsilon}\right]\right\} \tag{3.57}$$

As with the previous model, the Boussinesq approximation restricts the length of waves for which the theory is valid (Thorpe 1968).

The extension of the two-layer model, discussed in Chap. 2, is a multilayered thermocline model. In this model it is assumed that the pycnocline has a step-like structure made up of a series of layers (where density is almost constant) separated by thin sheets (where density changes sharply). According to Woods (1968) this is a good description of the summer thermocline of the Mediterranean Sea. Other specific cases of the density profiles are treated by Roberts (1975) in more detail.

For arbitrary density distributions, the vertical velocity equation (3.39) may be solved numerically using arbitrary values of $N(z)$ determined experimentally. In the early models, the Stömer or Runge-Kutta numerical methods were applied. However, in this monograph, for the three-layer type of density distribution and other density distributions, the variational calculations are applied in the next section.

3.5 Arbitrary Density Profile—Variational Approach

In the three-layer model discussed above, the adopted density non-uniform profile (3.37) resulted in a constant Väisälä-Brunt frequency N_0. In the general case this frequency is an arbitrary function of the vertical coordinate z. If we orient x-axis in the direction of the wave propagation, from Eq. (1.47) we obtain:

$$\frac{d^2W^{(n)}(z)}{dz^2} + \frac{N^2(z) - \omega^2}{\omega^2 - f^2} k_x^2 W^{(n)}(z) = 0 \tag{3.58}$$

Velocity $W^{(n)}(z)$ is assumed to be defined over an interval $[-h, 0]$ and we take the boundary conditions to be:

$$W^{(n)}(-h) = W^{(n)}(0) = 0 \tag{3.59}$$

The above boundary value problem should be valid for each mode, n. However, in the following the mode numbers have been omitted for simplicity.

Let us apply the approximate variational formulation for solving of the above boundary value problem (Kantorovitch and Krylov 1962). First, we represent Eq. (3.58) in a slightly different form:

$$L(W) = \frac{d^2W(z)}{dz^2} + q(z)\lambda W(z) = 0 \tag{3.60}$$

in which:

$$q(z) = \left(\frac{N^2(z) - \omega^2}{\omega^2 - f^2}\right), \quad \lambda = k_x^2 \tag{3.61}$$

Let the solution of Eq. (3.60) take form of the following summation:

$$W^{(n)}(z) = \sum_{j=1}^{m} a_j \varphi_j(z) \tag{3.62}$$

where:

$$\varphi_j(z) = \sin\left(\frac{j\pi z}{h}\right) \tag{3.63}$$

Coefficients a_j should be found from the condition of orthogonality of functions $\varphi_i(z)$ and functional on the left hand side of Eq. (3.60) after substitution of Eq. (3.62) for $W(z)$. Therefore, the orthogonality condition leads to the set of following equations:

$$\sum_{i=1}^{m}\left\{\int_{-h}^{0}\frac{d^2\varphi_j(z)}{dz^2}\,\varphi_i(z)dz+\lambda\int_{-h}^{0}q(z)\varphi_j(z)\varphi_i(z)dz\right\}a_j=0 \tag{3.64}$$
$$\text{for } i=1,2,\ldots,m \text{ and } j=1,2,\ldots,m$$

If we denote:

$$\alpha_{ij}=\int_{-h}^{0}\frac{d^2\varphi_j(z)}{dz^2}\,\varphi_i(z)dz \tag{3.65}$$

and

$$\gamma_{ij}=\int_{-h}^{0}q(z)\varphi_j(z)\varphi_i(z)dz \tag{3.66}$$

from Eq. (3.64) we have:

$$\sum_{j=1}^{m}\left(\alpha_{ij}+\lambda\gamma_{ij}\right)a_j=0 \quad \text{for } i=1,2,\ldots,m \tag{3.67}$$

Using Eq. (3.63) we define the coefficients α_{ij} as follows:

$$\alpha_{ij}=\begin{cases}-\dfrac{(j\pi)^2}{2h}, & i=j\\ 0, & i\neq j\end{cases} \tag{3.68}$$

Functions $q(z)$ are given by Eq. (3.61) and for three-layer density distribution (3.37), they are as follows:

$$q(z)=\begin{cases}\dfrac{-\omega^2}{\omega^2-f^2} & \text{for } -h_1\leq z\leq 0\\ \dfrac{N_0^2-\omega^2}{\omega^2-f^2} & \text{for } -h_2\leq z\leq -h_1\\ \dfrac{-\omega^2}{\omega^2-f^2} & \text{for } -h\leq z\leq h_2\end{cases} \tag{3.69}$$

in which Väisälä-Brunt frequency N_0 is given by Eq. (3.38).

System of homogeneous equations (3.67) has a nontrivial solution only when the determinant becomes equal to zero, i.e.:

$$\Delta = \begin{vmatrix} \alpha_{11} + \lambda\gamma_{11}, \alpha_{12} + \lambda\gamma_{12}, \ldots \alpha_{1n} + \lambda\gamma_{1n} \\ \alpha_{21} + \lambda\gamma_{21}, \alpha_{22} + \lambda\gamma_{22}, \ldots \alpha_{2n} + \lambda\gamma_{2n} \\ \ldots\ldots\ldots\ldots\ldots\ldots\ldots\ldots\ldots\ldots \\ \ldots\ldots\ldots\ldots\ldots\ldots\ldots\ldots\ldots\ldots \\ \alpha_{n1} + \lambda\gamma_{n1}, \alpha_{n2} + \lambda\gamma_{n2}, \ldots \alpha_{nn} + \lambda\gamma_{nn} \end{vmatrix} = 0 \tag{3.70}$$

Each eigenvalue λ, resulting from condition (3.70), corresponds to the particular horizontal wavenumber $k_x^{(n)}$ (see Eq. 3.61). Thus for a fixed f and ω for chosen eigenvalue $\lambda = k_x^{(n)}$, the system of homogeneous equations (3.67) yields the values of the coefficients a_{ij}, and finally the values of eigenfunction $W^{(n)}(z)$.

To illustrate the procedure of eigenvalues calculation let us consider the three-layer density distribution (3.37). The characteristic water depths and wave periods range were adopted as follows: $h_1 = 60\,\text{m}$, $h_2 = 200\,\text{m}$, $h = 240\,\text{m}$ and $0.25 < T < 1.5\,\text{h}$. The densities ρ_1 and ρ_2 are equal 1005.8 and 1009.0 kg m^{-3}, respectively. From Eq. (3.67) we determine the first three eigenvalues $\lambda_1, \lambda_2, \lambda_3$, and the first three wavenumbers $k_x^{(n)}$ as well as the phase velocities. They are shown in Fig. 3.10. When we compare the phase velocities in Figs. 3.5 and 3.10, we find that they are almost identical! The reason for this equivalency is the fact that the corresponding eigenvalues λ (or wavenumbers k) for both methods are almost the same. In fact they are different by less than 0.5 %.

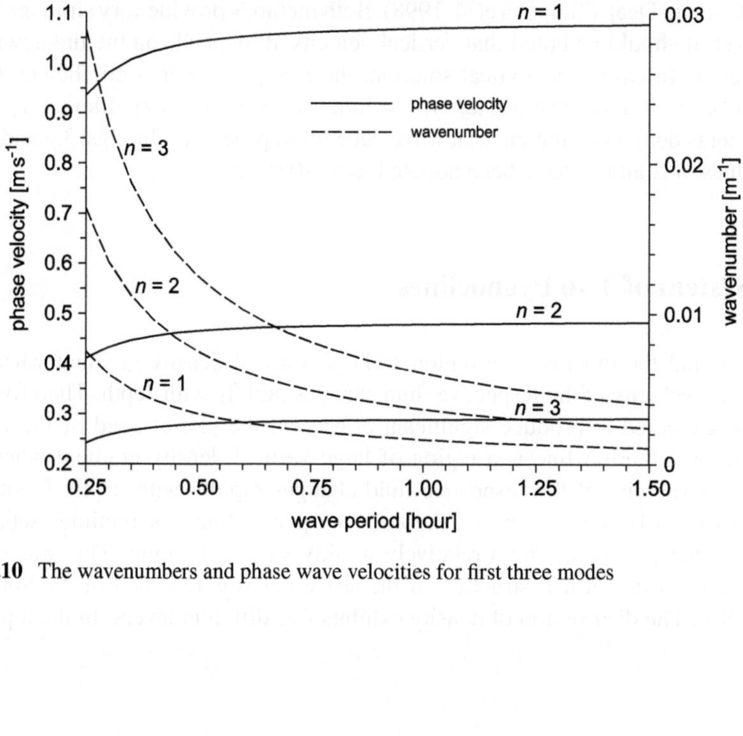

Fig. 3.10 The wavenumbers and phase wave velocities for first three modes

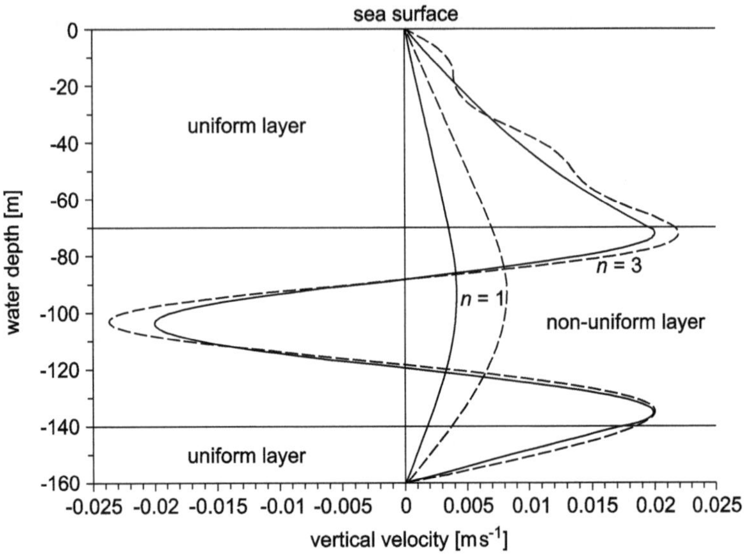

Fig. 3.11 Profiles of vertical velocity amplitudes for wave of period $T = 0.5$ h: analytical solution (3.42) (*solid line*) and variational type solution (3.62) (*broken line*)

In Fig. 3.11, the vertical profiles of velocity amplitudes $W^{(n)}$ are compared with velocity amplitudes resulted from analytical solution (3.42) for modes $n = 1$ and $n = 3$. The bathymetry as well as the water density data correspond to the Station 16 in the Gotland Deep (Talipova et al. 1998). Both methods provide very similar results. However, it should be noted that vertical velocity W depends on the unknown free parameters. In case of analytical solution, the free parameter is coefficient A (see Eq. 3.42), and for the variational type solution (3.62), this is coefficient a_1. Both coefficients depend on the internal wave generation processes. In Figs. 3.6 and 3.11, the values of A and a_1 have been adopted as ≈ -0.01 m s^{-1}.

3.6 System of Two Pycnoclines

The seasonal thermocline is a region of large vertical density gradient where the horizontal velocity of the respective fluid changes rapidly with depth. Therefore, the shear is expected to produce significant effects on the phase speed of the waves. The seasonal thermocline is a region of large vertical density gradient where the horizontal velocity of the respective fluid changes rapidly with depth. In shallow seas, particularly in the summer, the seasonal pycnocline is sometimes separated from another pycnocline by a relatively weakly stratified region. This was shown in Fig. 3.12. This vertical structure of the sea water was recorded in the Southern Baltic Sea. The distribution of density exhibits five different layers. In the top layer

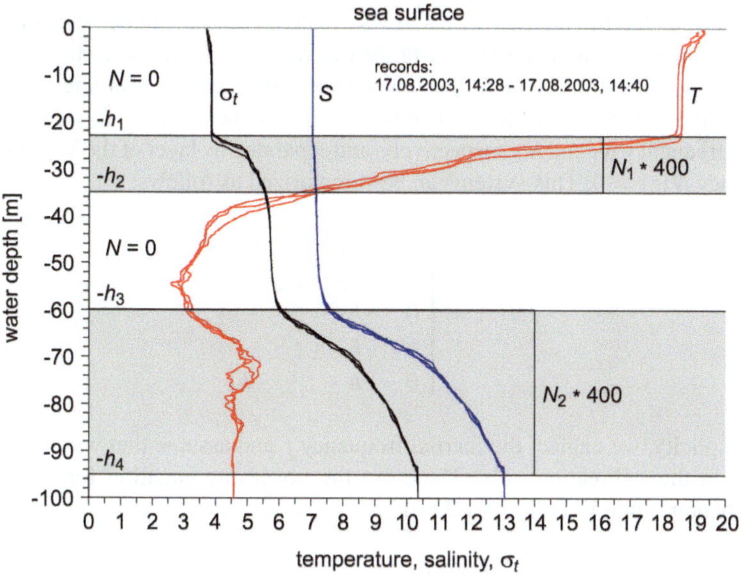

Fig. 3.12 Two-pycnocline system

($-h_1 \leq z \leq 0$), the density is constant $\rho = \rho_1$, thus the Väisälä-Brunt frequency $N(z) = 0$. The second layer shows a strong stratification which can be approximated for simplicity by the exponential law as:

$$\rho(z) = \rho_1 \exp\left(-\frac{N_1^2}{g} z\right) \tag{3.71}$$

in which frequency N_1 is:

$$N_1^2 = \frac{g \ln\left(\dfrac{\rho_1}{\rho_2}\right)}{h_1 - h_2} = \text{const} \tag{3.72}$$

where ρ_2 is the density at water depth $z = -h_2$.

Below the layer of strong stratification, the thick layer ($h_3 - h_2$) of very weak stratification is observed where frequency $N(z) = 0$. This layer separates the lower pycnocline from the upper seasonal pycnocline. We assume that within this lower pycnocline the density also behaves according the exponential law. Thus, frequency $N_2 = \text{const}$ for water depths ($-h_3 < z < -h_4$). Finally, in the bottom layer ($-h \leq z \leq -h_4$), density $\rho = \rho_3$ and frequency $N(z) = 0$.

Assuming that the internal waves are generated within the upper stratified layer, the question arises on the possibility of the transmission of the wave energy through the weak stratification layer to the lower pycnocline. Let us examine two lowest wave modes transferring energy between ducts of thicknesses $(h_2 - h_1)$ and $(h_4 - h_3)$ having frequency N_1 and N_2, respectively, and separated by layer of the Väisälä-Brunt frequency $N(z) = 0$. This system can be summarized as follows:

$$N(z) = \begin{cases} 0 & -h_1 \leq z \leq 0 \\ N_1 & -h_2 \leq z \leq -h_1 \\ 0 & -h_3 \leq z \leq -h_2 \\ N_2 & -h_4 \leq z \leq -h_3 \\ 0 & -h \leq z \leq -h_4 \end{cases} \tag{3.73}$$

For simplicity, we neglect the inertial frequency f and assume that waves are propagated in the x-direction only. Therefore the governing equation for the vertical velocity amplitude W is (see Eq. 3.60):

$$\frac{d^2W(z)}{dz^2} + q(z)\lambda\, W(z) = 0 \tag{3.74}$$

where:

$$q(z) = \frac{N^2(z) - \omega^2}{\omega^2}, \quad \lambda = k_x^2 \tag{3.75}$$

with the vertical motion restricted at the upper and lower boundaries, i.e. $W(0) = W(-h) = 0$. The given boundary value problem for W can be solved by the variational method discussed in Sect. 3.5. The characteristic water depths are shown in Fig. 3.12, and particular water densities are: $\rho_1 = 1004\,\mathrm{kg\,m^{-3}}$, $\rho_2 = 1006\,\mathrm{kg\,m^{-3}}$, $\rho_3 = 1010.4\,\mathrm{kg\,m^{-3}}$. The resulting values of the Väisälä-Brunt frequency are: $N_1 = 0.040326\,\mathrm{rad\,s^{-1}}$ and $N_2 = 0.034968\,\mathrm{rad\,s^{-1}}$. In Fig. 3.13, the dispersion relation for incident wave periods $5\,\mathrm{min} \leq T \leq 1\,\mathrm{h}$ is shown for two first modes. The corresponding phase velocity is given in Fig. 3.14.

To show the transference of energy from one strongly stratified region (h_1, h_2) through a weakly stratified region (h_2, h_3) to another strongly stratified region (h_3, h_4), we select two incident wave periods $T = 10\,\mathrm{min}$ and $T = 1\,\mathrm{h}$. Figures 3.15 and 3.16 illustrate the vertical distribution of the isopycnal displacements for two lowest modes. Patterns of isopycnals are similar as both wave periods are much longer than the buoyancy period $2\pi/N_1$. For the first mode, the isopycnal displacement is largest within the lower duct. However, for the second mode $(n = 2)$, the largest value of the isopycnal displacement is located within the upper duct. Therefore, the internal wave energy transfers through the very weak stratified layer from one duct to another.

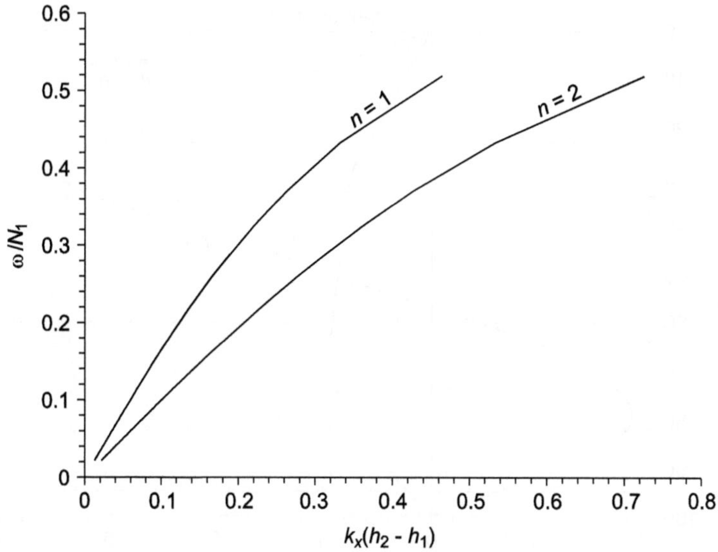

Fig. 3.13 Dispersion relation for incident wave period 5 min $\leq T \leq$ 1 h

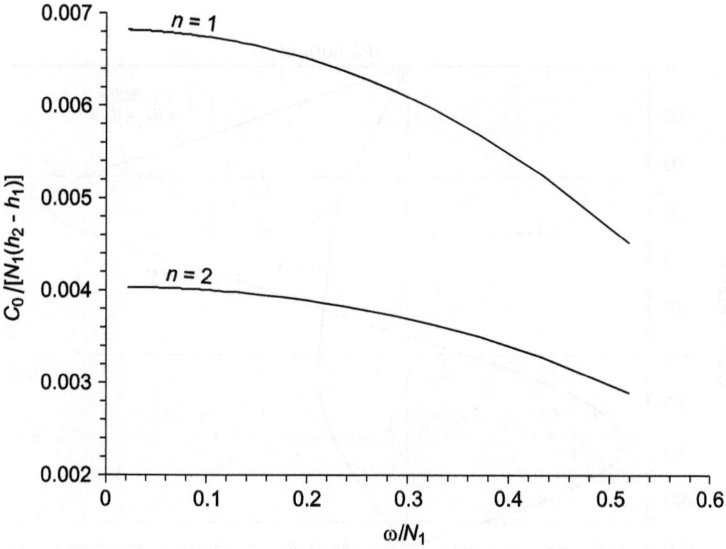

Fig. 3.14 Phase velocity for incident wave period 5 min $\leq T \leq$ 1 h

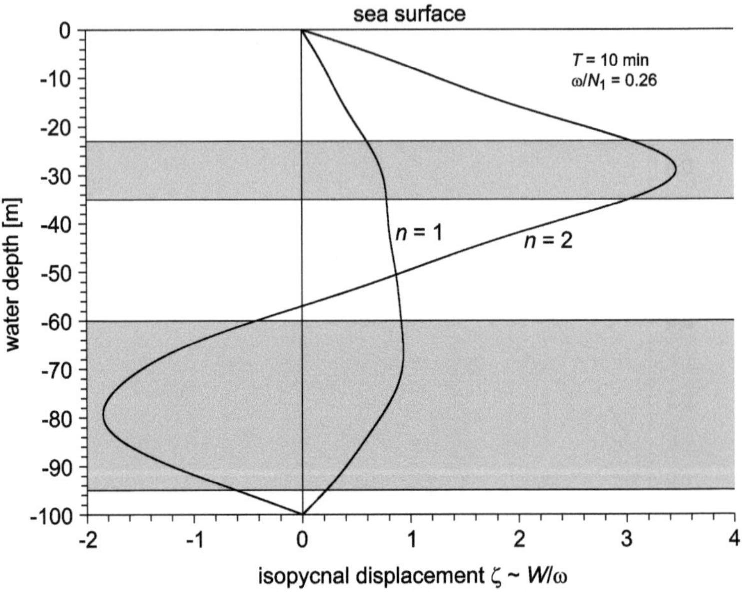

Fig. 3.15 Vertical distribution of isopycnal displacements for two lowest modes (wave period $T = 10\,\text{min}$)

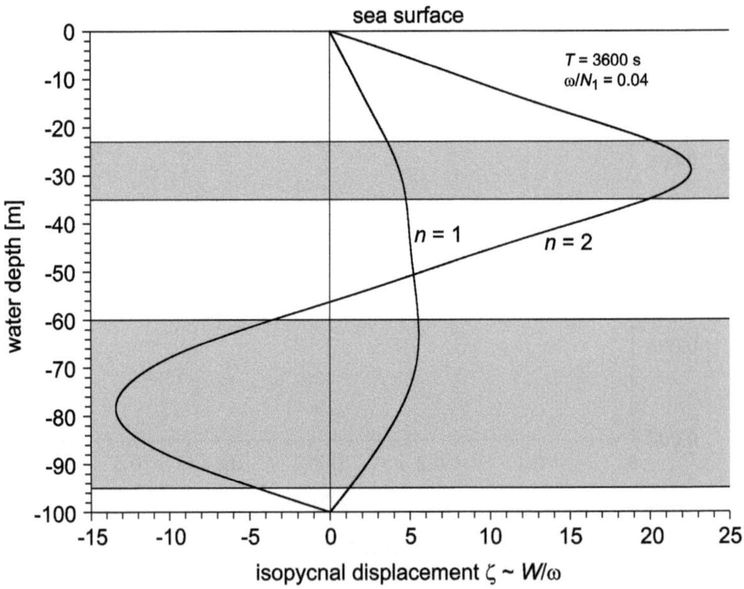

Fig. 3.16 Vertical distribution of isopycnal displacements for two lowest modes (wave period $T = 1\,\text{h}$)

3.7 Weakly Nonlinear Long Waves

The linear calculations performed so far for the internal waves propagating in the three-layer fluid give us a possibility to develop analytical solutions for a broad range of density distributions using the variational methodology. However, these solutions are strictly valid only in the limit of infinitesimally small amplitude waves. In shallow waters, the amplitudes of internal waves are usually finite, but not too large, and the wavelengths are relatively long compared to the fluid depth h. This introduces the weakly nonlinear effects to the wave field.

Similarly to the interfacial long waves (see Chap. 2), the theoretical descriptions of weakly nonlinear long internal waves in a fluid of continuous density stratification is based on the Kortweg-de Vries (KdV) equation (Pelinovsky et al. 1994, 2007; Grimshaw et al. 2004, 2007). Here we consider the long internal waves propagating in an arbitrary stratified fluid of constant water depth. Firstly, we present the vertical displacements of the isopycnal $\zeta(x, z, t)$ in the form of a product of two functions (Grimshaw et al. 2004):

$$\zeta(x, z, t) = \eta(x, t)\,\Phi(z) + \eta^2(z)T(z) \tag{3.76}$$

in which $\eta(x, t)$ describes the evolution of the internal wave profile in space and time, and functions $\Phi(z)$ and $T(z)$ determine the modal structure of the internal waves and its nonlinear correction.

For the periodic long waves when $\omega^2 \ll N^2(z)$ and the Earth rotation is neglected, we have:

$$N^2(z) - \omega^2 \approx N^2(z)\,, \quad \omega^2 - f^2 \approx \omega^2 \tag{3.77}$$

Therefore, Eq. (3.60) for vertical structure of internal waves $\Phi(z)$ becomes:

$$\frac{d^2\Phi(z)}{dz^2} + q(z)\lambda\Phi(z) = 0 \tag{3.78}$$

where:

$$q(z) = N^2(z)\,, \quad \lambda = \frac{k^2}{\omega^2} = \frac{1}{C_0^2} \tag{3.79}$$

Let us assume for simplicity that the nonlinear term in Eq. (3.76) is omitted. Thus, the long internal waves evolution in space and time is described by the classical Korteweg-de Vries equation:

$$\frac{\partial \eta}{\partial t} + C_0\frac{\partial \eta}{\partial x} + \alpha\eta\frac{\partial \eta}{\partial x} + \beta\frac{\partial^3 \eta}{\partial x^3} = 0 \tag{3.80}$$

in which the nonlinear coefficient α and dispersion parameter β are given by (Pelinovsky et al. 2007):

$$\alpha = \frac{3}{2} C_0 \frac{\int\limits_{-h}^{0} \left(\frac{d\Phi(z)}{dz} \right)^3 dz}{\int\limits_{-h}^{0} \left(\frac{d\Phi(z)}{dz} \right)^2 dz} \tag{3.81}$$

and

$$\beta = \frac{1}{2} C_0 \frac{\int\limits_{-h}^{0} \Phi^2(z)\, dz}{\int\limits_{-h}^{0} \left(\frac{d\Phi(z)}{dz} \right)^2 dz} \tag{3.82}$$

To proceed further we assume that the density profile is arbitrary. Therefore, the Väisälä-Brunt frequency $N(z)$ is also an arbitrary function of coordinate z. To get the solution of the eigenvalue problem for function $\Phi(z)$ we apply the variational approach discussed in the Sect. 3.5 for particular mode n as follows:

$$\Phi^{(n)}(z) = \sum_{j=1}^{M} a_j \varphi_j(z) \tag{3.83}$$

where function $\varphi_j(z)$ is given by Eq. (3.63).

When the parameters α and β are known, the solution of Korteweg-de Vries equation (3.80) has a form of periodic waves. In particular, namely the cnoidal waves or solitary waves. In particular, for the cnoidal type solution we obtain:

$$\eta(x,t) = \frac{H}{m} \left(1 - \frac{\mathbf{E}}{\mathbf{K}} - m \right) + Hcn^2 \left[\frac{2\mathbf{K}(m)}{L} (x - Ut) \right] \tag{3.84}$$

in which H is the wave height, L is the wavelength and $cn(x)$ is the Jacobian elliptic function (Abramowitz and Stegun 1975). For later convenience we rewrite wavelength L in the form of a transcendental relationship for unknown parameter m:

$$\frac{\hat{\alpha} H}{48} \left(\frac{L}{\sqrt{\hat{\beta}}} \right)^2 = m\mathbf{K}^2 \tag{3.85}$$

where:

$$\hat{\alpha} = \frac{\alpha}{C_0} \quad \text{and} \quad \hat{\beta} = \frac{\beta}{C_0} \tag{3.86}$$

The cnoidal wave speed is given in terms of ζ_1, ζ_2, and ζ_3 (see Eqs. 2.71–2.73):

$$\frac{U}{C_0} = 1 + \frac{\hat{\alpha}}{3}(\zeta_1 + \zeta_2 + \zeta_3) \tag{3.87}$$

The second solution of Eq. (3.80), i.e. the solitary wave is given by:

$$\eta(x,t) = A\cosh^{-2}\left[\left(\frac{\hat{\alpha}A}{12\hat{\beta}}\right)^{1/2}(x - Ut)\right] \tag{3.88}$$

and the solitary wave speed becomes:

$$U = C_0\left(1 + \frac{\hat{\alpha}}{3}A\right) \tag{3.89}$$

The velocities generated by internal waves can be calculated as follows (Holloway et al. 1999; Kurkina et al. 2011):
– vertical velocity $w(x,z,t)$

$$w(x,z,t) = \frac{\partial\zeta(x,z,t)}{\partial t} = \Phi(z)\,\frac{\partial\eta(x,t)}{\partial t} \tag{3.90}$$

or

$$w(x,z,t) = C_0\Phi(z)\,\frac{\partial\eta(x,t)}{\partial x} \tag{3.91}$$

– horizontal velocity

$$u(x,z,t) = -C_0\,\frac{d\Phi(z)}{dz}\,\eta(x,t) \tag{3.92}$$

Therefore, in the case of periodic pycnoclines for the cnoidal waves we obtain:

$$w(x,z,t) = -4C_0\left(\frac{H}{L}\right)\mathbf{K}(m)\,sn(x_1,m)\,cn(x_1,m)\,dn(x_1,m)\,\Phi(z) \tag{3.93}$$

and

$$u(x,z,t) = -C_0H\left[\frac{1}{m}\left(1 - \frac{\mathbf{E}}{\mathbf{K}} - m\right) + cn^2(x_1,m)\right]\frac{d\Phi(z)}{dz} \tag{3.94}$$

where:

$$x_1 = \frac{2\mathbf{K}(m)}{L}(x - Ut) \tag{3.95}$$

The functions $sn(x_1, m)$, $cn(x_1, m)$ and $dn(x_1, m)$ are Jacobian elliptic functions (Abramowitz and Stegun 1975) and the following relationships between them are valid:

$$\left.\begin{aligned} sn^2(x_1, m) + cn^2(x_1, m) = 1 \\ dn^2(x_1, m) = 1 - m\, sn^2(x_1, m) \end{aligned}\right\} \tag{3.96}$$

When we assume that pycnoclines have a form of solitary waves, the velocities w and u are:

$$w(x, z, t) = -2AC_0 \left(\frac{\hat{\alpha}A}{12\hat{\beta}}\right)^{1/2} \cosh^{-3}(x_1) \sinh(x_1)\, \Phi(z) \tag{3.97}$$

$$u(x, z, t) = -AC_0 \cosh^{-2}(x_1) \frac{d\Phi(z)}{dz} \tag{3.98}$$

where:

$$x_1 = \left(\frac{\hat{\alpha}A}{12\hat{\beta}}\right)^{1/2} (x - Ut) \tag{3.99}$$

To get the density structure induced by internal waves we refer to Eq. (1.24), i.e.:

$$\rho(x, z, t) = -\zeta(x, z, t) \frac{d\bar{\rho}(z)}{dz} \tag{3.100}$$

and total density $\rho^*(x, z, t)$ becomes:

$$\rho^*(x, z, t) = \bar{\rho}(z) + \rho(x, z, t) = \bar{\rho}(z) - \zeta(x, z, t) \frac{d\bar{\rho}(z)}{dz} \tag{3.101}$$

Let us illustrate the application of the methodology, described above for the vertical profiles of the temperature T, salinity S, density σ_t and Väisälä-Brunt frequency N recorded on 5.02.2003, 6:17 at the point (55°15.084 N, 17°25.623 E)—see Fig. 1.1.

In Fig. 3.17, pycnoclines induced by long cnoidal type internal wave are given. The first mode was assumed ($n = 1$), with wave period $T = 0.5$ h, wavelength $L = 1.432$ km, and wave height $H = 5$ m. Note that the cnoidal type evolution of pycnoclines due to internal waves is clearly seen only within the non-uniform water depth layer (~50–73 m).

Figure 3.18 shows the vertical profiles of velocity components under wave crest and wave through. Vertical velocity components at these profiles of course equal zero. However, the horizontal components reach their maximum value, and change their sign at water depth where the vertical gradient of function $\Phi(z)$ equals zero.

Figure 3.19 illustrates the vertical profiles of velocity components at distances x_1 where vertical velocity becomes maximal, i.e. when product $sn(x_1, m) \cdot cn(x_1, m) \cdot dn(x_1, m) \rightarrow \max$. At the rising slope of sea surface, the vertical velocity is positive

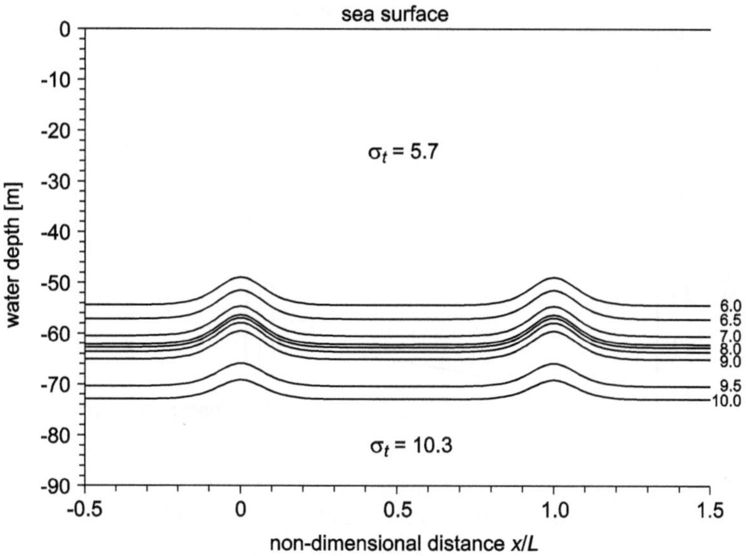

Fig. 3.17 Pycnocline structure for cnoidal type internal waves of period $T = 0.5\,\text{h}$

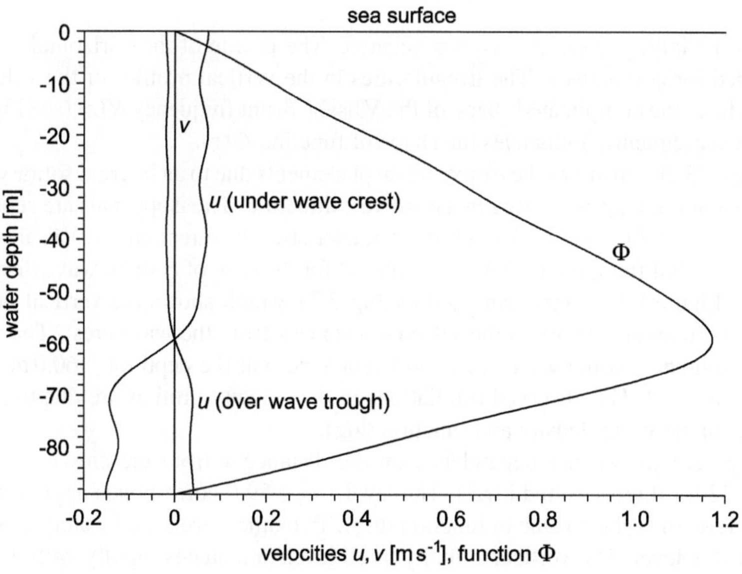

Fig. 3.18 Vertical profiles of velocity components under crest and trough of cnoidal waves

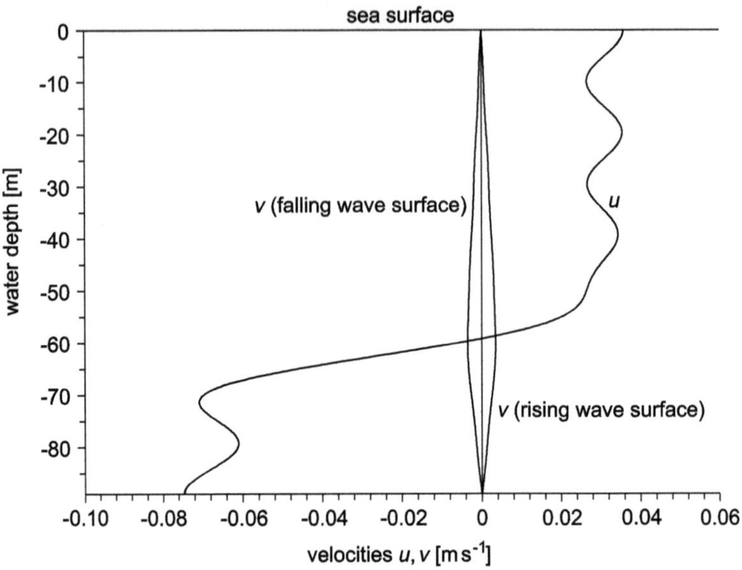

Fig. 3.19 Vertical profiles of cnoidal wave components at point where vertical velocity becomes maximal

and at the falling slope, it becomes negative. The profile of the horizontal velocity is added for comparison. The irregularities in the vertical profiles of the velocities result from the complicated shape of the Väisälä-Brunt frequency $N(z)$ (see Fig. 1.1) which subsequently influences the shape of function $\Phi(z)$.

Figure 3.20 illustrates the isopycnal displacements due to an internal solitary wave of 5 m amplitude, propagating in a positive x direction. The isopycnals are separated in the vertical direction by 5 m. Other characteristics of water density, Väisälä-Brunt frequency and function $\Phi(z)$ are the same as for the case of cnoidal wave discussed above. Figure 3.20 is supplemented by Fig. 3.21 which shows the vertical profiles of the horizontal velocity at the selected distances from the wave crest. The velocities are highest at the wave crest and become zero at the depth of $\approx$60.0 m where $d\Phi(z)/dz = 0$. The observed oscillations in the velocity profiles are due to irregularities in the water density and function $\Phi(z)$.

The vertical velocity dependence on the distance x from the crest is given in Fig. 3.22 for three selected levels. The level $z = -59.0$ corresponds approximately to the level of highest value of function $\Phi(z)$. Two other levels are located above and below this level. The vertical velocity at all levels attenuates rapidly with distance from wave crest.

For coastal waters the coefficient of the nonlinear term α in the KdV equation sometimes becomes very small. This is the case when the pycnocline lies near the middle depth of fluid. Then the higher-order correction terms should be taken into account in the KdV equation. Such equation is known as extended Korteweg-de Vries equation (eKdV). For continuous density stratification and some background shear flow $U(z)$ the (eKdV) equation takes the form (Grimshaw et al. 2004):

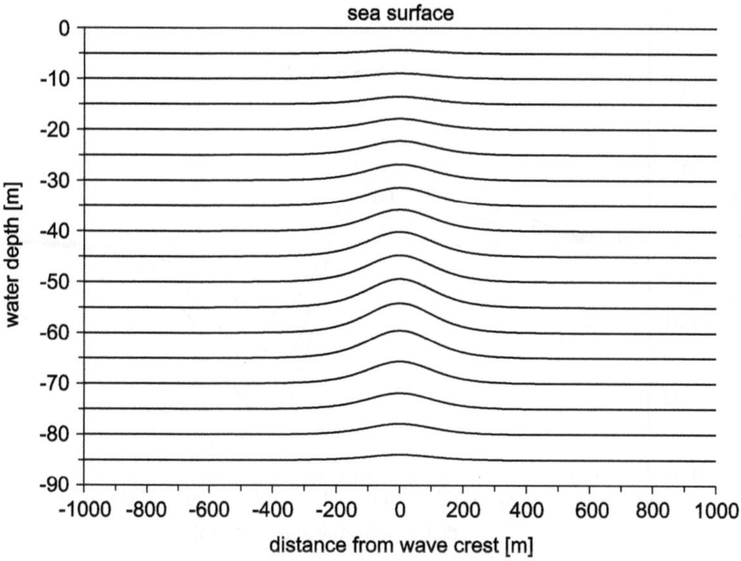

Fig. 3.20 Isopycnal displacements due to solitary wave of amplitude of 5 m

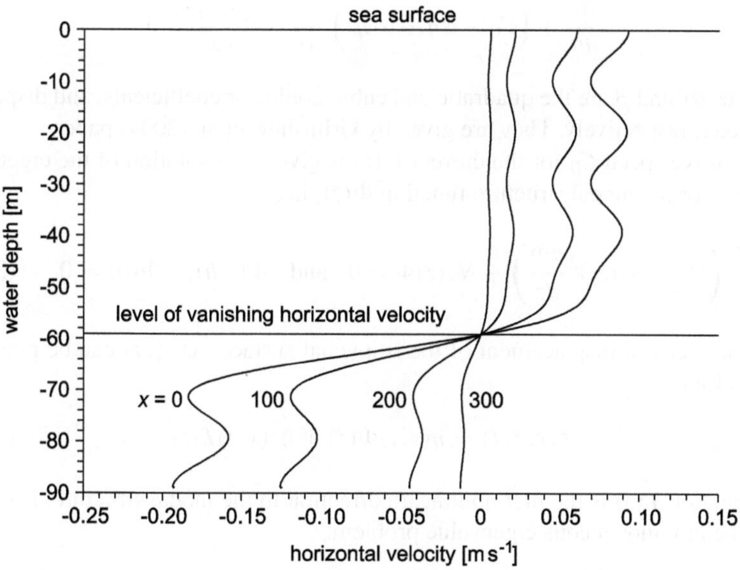

Fig. 3.21 Vertical profiles of horizontal velocity of a solitary wave at the selected distances from the wave crest

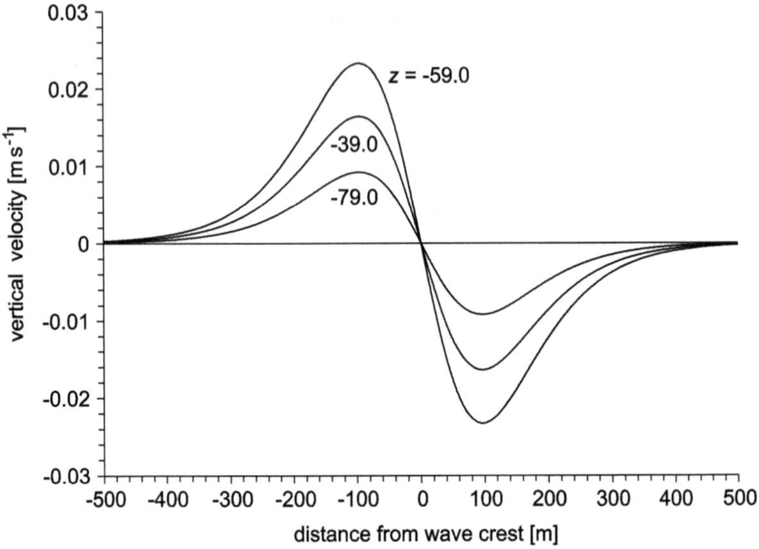

Fig. 3.22 Vertical velocity of solitary waves as a function of distance for three selected levels

$$\frac{\partial \eta}{\partial t} + \left(C_0 + \alpha \eta + \alpha_1 \eta^2\right) \frac{\partial \eta}{\partial x} + \beta \frac{\partial^3 \eta}{\partial x^3} = 0 \tag{3.102}$$

where α, α_1 and β are the quadratic and cubic nonlinear coefficients, and dispersion parameter, respectively. They are given by Grimshaw et al. (2004) paper.

The wave speed C_0 for the current $U(z)$ is given as a solution of the eigenvalue problem for the modal structure function $\Phi(z)$, i.e.:

$$\frac{d}{dz}\left([C_0 - U(z)]^2 \frac{d\Phi}{dz}\right) + N^2(z)\Phi = 0 \quad \text{and} \quad \Phi(-h) = \Phi(0) = 0 \tag{3.103}$$

The final vertical displacement of the isopycnal surface $\zeta(x, z, t)$ can be presented now as follows:

$$\zeta(x, z, t) = \eta(x, t)\Phi(z) + \eta^2(x, t)T(z) \tag{3.104}$$

The function $T(z)$ is the first nonlinear correction to the modal structure. It follows from the inhomogeneous eigenvalue problem:

$$\frac{d}{dz}\left[(C_0 - U)^2 \frac{dT}{dz}\right] + N^2 T = -\alpha \frac{d}{dz}\left[(C_0 - U)\frac{d\Phi}{dz}\right] + \frac{3}{2}\frac{d}{dz}\left[(C_0 - U)^2 \left(\frac{d\Phi}{dz}\right)^2\right] \tag{3.105}$$

with zero boundary conditions on the sea surface and sea bottom.

Grimshaw et al. (2003) employed the eKdV equation, supplemented with a damping term to study the damping of the solitary waves. The damping term included several different forms of friction, using both the analytical adiabatic asymptotic theory and numerical simulations. They found that in the case of linear friction the damping of large-amplitude solitary wave induces the formation of a tail of the same polarity as the solitary wave. On the other hand, when the dissipation is not too small, the tail formed behind the large solitary wave transforms into a group of secondary waves which is separated from the primary waves. In general, the features of the solitary waves damping are closely depending on the sign of the cubic nonlinear term and linear dispersion term.

3.8 Wave Energy in a Stratified Fluid

The classical formula for the total potential energy is resulted from the volume integral of $\rho g z$. For example, in the case of surface waves, the potential (E_p) and kinetic (E_k) energies (averaged for unit sea surface) are expressed in the form of wave height H as follows (Massel 2013):

$$E_p = E_k = \rho_w g \frac{H^2}{16} \tag{3.106}$$

From this definition it follows that the potential energy is converted entirely into kinetic energy only if the fluid as a whole is brought down to some reference level z, say $z = 0$. However, not all potential energy can be converted into kinetic energy, which finally contributes to mixing and heat. For the internal waves, the portion of the potential energy available for this conversion is known as the available potential energy (APE). To develop the formula for the available potential energy let us assume that energy is distributed in the whole continuously stratified water column and for simplicity we neglect the Earth rotation. Following Kundu and Cohen (2006) we multiply Eq. (1.31) by $\overline{\rho}(z)u$, Eq. (1.32) by $\overline{\rho}(z)v$ and Eq. (1.33) by $\overline{\rho}(z)w$, and summarize them to get:

$$\frac{\partial}{\partial t}\left[\frac{1}{2}\overline{\rho}(z)\left(u^2 + v^2 + w^2\right)\right] + g\rho(z)w + \nabla \cdot (p\mathbf{u}) = 0 \tag{3.107}$$

The first term in Eq. (3.107) represents the rate of change of the kinetic energy. Therefore, the second term can be considered as the rate of change of potential energy. It should be noted that the last term is the net work done by the pressure forces or it can be interpreted as the divergence of the energy flux $p\mathbf{u}$. This flux induces energy changes at a given point in the fluid column. Thus, for potential energy we get:

$$\frac{\partial E_p}{\partial t} = g\rho(z)\,w(z) \tag{3.108}$$

When Eq. (1.22) is used in the expression above we obtain:

$$\frac{\partial E_p}{\partial t} = \frac{\partial}{\partial t}\left[\frac{g^2\rho^2(z)}{2\overline{\rho}(z)N^2(z)}\right] \tag{3.109}$$

and

$$E_p(z) = \frac{1}{2}\frac{g^2\rho^2(z)}{\bar{\rho}(z)N^2(z)} \quad \left[\frac{\mathrm{J}}{\mathrm{m}^3}\right] \tag{3.110}$$

which is the potential energy for linear waves per unit volume at a given level z.

As was shown in Sect. 3.3, the governing equations for internal wave motion can also be expressed in terms of the vertical velocity w. Thus, using Eqs. (3.46) and (3.52), we can represent the potential energy for periodic internal waves of frequency ω, propagating in vertical plane (x, z) as follows:

$$E_p(z) = \frac{g^2}{2\omega^2}\frac{W^2(z)}{\overline{\rho}(z)N^2(z)}\left(\frac{\partial\overline{\rho}(z)}{\partial z}\right)^2\left(\sin^2(kx-\omega t)\right) \tag{3.111}$$

where $\rho(x, z, t) \sim \exp[i(kx - \omega t)]$ and $w(x, z, t) \sim \exp[i(kx - \omega t)]$.

If we average the expression above over a wave period and assume $x = 0$ we obtain:

$$\overline{E_p}(z) = \frac{g^2}{4\omega^2}\frac{W^2(z)}{\overline{\rho}(z)N^2(z)}\left(\frac{\partial\overline{\rho}}{\partial z}\right)^2 \quad \left[\frac{\mathrm{J}}{\mathrm{m}^3}\right] \tag{3.112}$$

The potential energy for stratified fluid should be considered as energy averaged per wave period and per unit volume. In Eq. (3.112) we used the fact that the average of $\sin^2(\omega t)$ over wave period equals 1/2.

Taking into account Eq. (1.7) we rewrite Eq. (3.112) as follows:

$$\overline{E_p}(z) = \frac{\overline{\rho}(z)}{4}\left(\frac{N(z)W(z)}{\omega}\right)^2 \quad \left[\frac{\mathrm{J}}{\mathrm{m}^3}\right] \tag{3.113}$$

In the particular case of the exponential density profile (3.1), the amplitude of vertical velocity $W(z)$ is given by Eq. (3.15). After substituting amplitude $W(z)$ into Eq. (3.113), the average potential energy per unit mass at a given level z becomes:

$$\overline{E_p}(z) = \frac{N^2(z)\,\overline{\rho}(z)}{4\omega^2}B_n^2\sin^2(k_z z) \tag{3.114}$$

From the physical point of view, it is convenient to express the potential energy as a function of the displacement ζ of fluid particle. From Eq. (1.24) we have:

$$\rho(z,t) = -\zeta(z,t)\,\frac{\partial\overline{\zeta}(z)}{\partial z} \tag{3.115}$$

or

$$\rho(z,t) = \frac{N^2(z)\,\overline{\rho}(z)}{g}\,\zeta(z,t) \tag{3.116}$$

After substituting Eq. (3.116) into Eq. (3.110), we obtain the averaged potential energy in the form:

$$\overline{E_p}(z) = \frac{1}{4}N^2(z)\,\overline{\rho}(z)\,\overline{\zeta^2}(z) \quad \left[\frac{\mathrm{J}}{\mathrm{m}^3}\right] \tag{3.117}$$

It should be noted that the available potential energy, given by Eqs. (3.113) and Eq. (3.117), are related to the particular mode, n, being associated with the Boussinesq approximation for slowly varying stratification (Carter et al. 2008).

From Eq. (3.107) it follows that the kinetic energy per unit volume at a given level z is:

$$\overline{E_k}(z) = \frac{1}{2}\overline{\rho}(z)\overline{\left(u^2(z) + w^2(z)\right)} \tag{3.118}$$

In particular, for an exponential density profile, the kinetic energy, averaged over wave period and unit mass, equals:

$$\overline{E_k}(z) = \frac{1}{4}\,\overline{\rho}(z)\left[\left(\frac{k_z}{k_x}\right)^2 \cos^2(k_z z) + \sin^2(k_z z)\right]B_n^2 \tag{3.119}$$

or

$$\overline{E_k}(z) \approx \frac{1}{4}\,\overline{\rho}(z)\left(\frac{N^2(z)}{\omega^2}\right)\cos^2(k_z z)B_n^2 \tag{3.120}$$

where $k_z = n\pi/h$. The expression above can be written as follows:

$$\overline{E_k}(z) \approx \frac{1}{4}\,\overline{\rho}(z)N^2(z)\,\overline{\zeta^2}(z) \tag{3.121}$$

When the vertical variation of the water density is negligibly small, i.e. $\overline{\rho}(z) \approx \rho_0$, we can represent the averaged potential energy of the entire water column per unit horizontal area as follows:

$$\langle E_p \rangle = \frac{1}{h}\int_{-h}^{0} \overline{E_p}(z)\,dz \tag{3.122}$$

After substituting Eq. (3.114) we obtain:

$$\langle E_p \rangle = \frac{N_0^2 \rho_0}{8\omega^2} B_n^2 = \frac{1}{8} \rho_0 N_0^2 \overline{A}_\zeta^2 \tag{3.123}$$

where A_ζ is the vertical displacement amplitude.

Similarly from Eq. (3.121) we have:

$$\langle E_k \rangle = \frac{N_0^2 \rho_0}{8\omega^2} B_n^2 = \frac{1}{8} \rho_0 N_0^2 \overline{A}_\zeta^2 \tag{3.124}$$

Therefore, under the assumption of a very slowly varying density, both vertically averaged energies are the same and total internal wave energy becomes:

$$\langle E \rangle = \overline{E_p} + \overline{E_k} = \frac{\rho_0 N_0^2}{4} \overline{A}_\zeta^2 \tag{3.125}$$

It should be noted that the average potential energy of the entire water column per unit horizontal area for interfacial waves in infinitely deep fluid is (Kundu and Cohen 2006):

$$\overline{E_p} = \frac{1}{8} (\rho_2 - \rho_1) g A_\zeta^2 \tag{3.126}$$

where the interface displacement $\zeta = A_\zeta \cos(kx - \omega t)$ and $(\rho_2 - \rho_1)$ is the density discontinuity.

Another method of calculation of the available potential energy is based on the calculation of difference between potential energy of the perturbed and unperturbed states. The details of calculation are given by Klymak et al. (2006), Moum et al. (2007), Lamb and Nguyen (2009), Kang and Fringer (2010) and Lien et al. (2014). There are a few definitions of APE discussed by Kang and Fringer (2010). The first definition, $\text{APE}_1 = \rho g z$, is used to calculate the depth-integrated or domain-integrated APE. According the second definition, an exact local APE_2, for arbitrary stratification takes the form (Holliday and McIntyre 1981):

$$\text{APE}_2 = \int_{z-\zeta}^{z} g[\overline{\rho}(z) - \rho_r(z')]dz' \tag{3.127}$$

where ρ_r is the reference density. This formulation was employed more recently in analysing the energetics of nonlinear internal waves (Lien et al. 2014). Comparison of three different APE formulations and their performance in numerical simulations of progressive nonlinear internal waves showed that APE_2 in slightly varying stratification is the most appropriate for evaluating the local APE.

3.9 Remarks on Wave Induced Mixing in a Stratified Fluid

3.9.1 Mixing in Two-Layer Flow System

To mix a stably stratified fluid, energy is required. The turbulent kinetic energy required to generate and maintain mixed layers is usually produced by bottom friction, large-scale currents, shoaling, reflection of high frequency internal waves and interaction of currents with rough topography. Mixing by short internal waves and turbulence generates vertical motions and overturning. In a stratified fluid, the work must be done to raise heavy fluid elements and lower light elements. Following Cushman-Roisin (1994), let us consider a two-layer system with upper horizontal layer of thickness h_1 and density ρ_1, and lower horizontal layer of thickness $h_2 = h_1$ and density $\rho_2 > \rho_1$. The average velocities in the layers are U_1 and U_2, respectively. For such system, initially the centre of gravity is located below the middle depth. However, in the final state, after complete mixing due to action of turbulence and internal waves, the centre of gravity will be located exactly at the middle depth. Thus, the potential energy, *PE*, is needed to raise the centre of gravity:

$$PE = \int_0^h \rho_f gz dz - \int_0^h \rho_i gz dz = \frac{1}{8}(\rho_2 - \rho_1) g h^2 \tag{3.128}$$

where $h = 2h_1 = 2h_2$, ρ_i are the initial densities different in both layers, and $\rho_f = (\rho_1 + \rho_2)/2$.

As the initial velocity distribution is uneven, the kinetic energy is released in the process in which, in the absence of external forces, the velocities in both layers will be finally the same, i.e. $U = (U_1 + U_2)/2$. Therefore, conservation of linear momentum gives:

$$KE_{\text{loss}} = \int_0^h \frac{1}{2}\rho_0 u_i^2 dz - \int_0^h \frac{1}{2}\rho_0 u_f^2 dz = \frac{1}{8}\rho_0 (U_1 - U_2)^2 h \tag{3.129}$$

where $\rho_0 \approx \rho_1 \approx \rho_2$, taking into account the Boussinesq approximation.

Now, the vertical mixing is possible as long as the kinetic energy loss exceeds the potential energy gain. Thus we have:

$$\frac{(\rho_2 - \rho_1) g h}{\rho_0 (U_1 - U_2)^2} < 1 \tag{3.130}$$

From this equation it follows that the complete mixing is possible when the initial density difference is sufficiently weak and the interface is not a strong gravitational barrier, or the initial velocity shear should be sufficiently large to supply the necessary amount of energy.

If the criterion (3.130) is not satisfied, mixing occurs only in the vicinity of the initial interface. This localised mixing is the same as the instability by interfacial waves of some wave lengths. According to Kundu and Cohen (2006), the sinusoidal perturbation of wavenumber k is unstable if:

$$\left(\rho_2^2 - \rho_1^2\right) g < \rho_1 \rho_2 k \left(U_1 - U_2\right)^2 \tag{3.131}$$

The flow perturbations extend on both side of the interface over a depth on the order of their wave length. For the breaking waves and the turbulent mixing induced by them, the vertical extent of the mixing zone Δh is proportional to the wave length of the longest unstable wave. Thus we have (Cushman-Roisin 1994):

$$\Delta h \approx \frac{\rho_0 (U_1 - U_2)^2}{2(\rho_2 - \rho_1) g} \tag{3.132}$$

3.9.2 Role of Internal Waves in Supporting Turbulent Mixing

Internal waves play an important role in the balance of stratification and supporting turbulent mixing. Since internal waves can be generated through the interaction of tides with the sloping underwater topography, the most substantial mixing occurs above such topography. The strong mixing is able to support the vertical turbulent diffusivity of the order of $10^{-4}\,\mathrm{m^2\,s^{-1}}$ which is needed to maintain the vertical density stratification in the ocean interior (Munk and Wunsch 1998).

Davies and Xing (2007) reported the results of the numerical modelling of the tidal mixing in the region of sills. A vertical, cross-sectional, non-hydrostatic model was used for the sill at the entrance to a Scottish Loch. By using an idealized topography, the effects of stratification and tidal velocity upon lee wave formation and associated vertical mixing have been examined. Enhanced mixing on the lee side was generated by density overturning and shear instabilities related to the hydraulic transition and the separation of the jet on that side. The mixing increased with increasing Froude number $F = U/(Nh_s)$, where U is the tidal velocity over the sill, N is the Väisälä-Brunt frequency and h_s is the still water depth. At high F values with strong currents, short internal waves and overturning occurred.

Beside the influence of the winds and bottom topography, the mixing is induced by internal waves breaking (Ivey and Nokes 1989). For a real fluid, the instability occurs when the wave amplitude exceeds the critical amplitude. Non-linear internal waves break when their Froude number $Fr = U/(NL)$ becomes greater than 1. Wave breaking produces motions which irreversibly modify their environment due to a transport of mass along the vertical direction through fluid mixing as well as along the horizontal direction through the modification of the momentum.

The growth of an instability upon a basic state leads to overturning of isopycnals when heavy fluid appears over light fluid. A primary wave becomes unstable against secondary waves acting as perturbations to the primary wave. The resonant mechanism between the three wave vectors and three frequencies is possible under the condition that $\omega_1 + \omega_2 + \omega_3 = 0$ and $\mathbf{k_1} + \mathbf{k_2} + \mathbf{k_3} = 0$. The energy of the perturbation components is transferred from the primary wave and when the steepness of the perturbation waves become of order 1, the isopycnals are locally overturned (Bouruet-Aubertot et al. 2001; Staquet and Sommeria 2002; Staquet 2004).

The universal character of the internal wave spectra suggests that they are close to a marginal saturation state. The corresponding vertical turbulent "diapycnal" exchange across the stratification is not very large but still it is about 100 times larger than molecular diffusion. This turbulence rate does not destroy stratification, but provides sufficient exchange to supply nutrients from the bottom to the euphotic zone in a shelf sea (van Haren 2012). Moreover, the mixing of cold water masses, which results in a slow upwelling motion of those masses (with speed of $\sim 10^{-7}\,\mathrm{m\,s^{-1}}$) and partly accounts for the thermal balance of the global ocean, is induced by the mixing of breaking internal waves of tidal origin (Stigebrandt 1979; Munk and Wunsch 1998). Hence, these local and short-time processes may appear to contribute fundamentally to the balance of the Earth's climate.

Let us now provide some more general remarks on mixing due to internal waves motion. We consider a fluid at rest in the form of two layers of different densities, with the denser layer being the lower layer and the density gradient being very smooth due to mass flux across the density interface due to diapycnal mass flux. Thus, the diffusion equation becomes (Staquet 2004):

$$\frac{\partial \rho}{\partial t} + \frac{\partial F_{\mathrm{mol}}(z)}{\partial z} = 0 \tag{3.133}$$

where $F_{\mathrm{mol}}(z)$ is the local molecular diapycnal mass flux:

$$F_{\mathrm{mol}}(z) = -\kappa \frac{d\rho}{dz} \tag{3.134}$$

where κ is the molecular diffusivity.

Due to the mass flux, the centre of mass will rise. Therefore, the potential energy of fluid E_p becomes:

$$E_p = \frac{1}{V} \int_V \rho(z) g z \, dV \tag{3.135}$$

in which V is the volume of fluid. Following Staquet (2004), we assume that the fluid volume has a form of fixed parallelepiped box, bounded by surface S. An evolution of potential energy is:

$$\frac{dE_p}{dt} = P_{\mathrm{mol}} + F_{\mathrm{surf}} \tag{3.136}$$

where P_{mol} is the rate of production of potential energy in the fluid at rest:

$$P_{mol} = \frac{g}{V} \int_V F_{mol} \, dV = \frac{\kappa g}{H} \Delta\rho \tag{3.137}$$

in which H is the height of fluid volume and $\Delta\rho = \rho(0) - \rho(H)$ is the positive density difference between the bottom and the top of the fluid container. The F_{surf} is the laminar rate of change of energy E_p due to diffusion across the surface S.

When fluid is moving with the speed $\mathbf{u}$, the mixing occurs due to local diapycnal diffusive flux of mass, F_{dyn}. To determine the value of F_{dyn} let us assume that in some moment the fluid relaxes instantaneously towards a rest state through an adiabatic redistribution of density, which results in horizontal isopycnals with stable density profile $\rho(z)$. Now, the potential energy is the minimum energy of the fluid according to Lorentz (1955). Due to the fact that the changes in density ρ are only due to diapycnal mass flux F_{dyn}, we obtain (Staquet 2004):

$$\frac{\partial \rho}{\partial t} + \frac{\partial F_{dyn}}{\partial z} = 0 \tag{3.138}$$

The solution of Eq. (3.138) takes a form (Winters and D'Asaro 1996):

$$F_{dyn}(z) = \kappa \frac{\langle |\nabla \rho|^2 \rangle}{-d\rho/dz} > 0 \tag{3.139}$$

where $\langle \ \rangle$ denotes an average value along isopycnals and ρ is the total density field.

Per analogy to the rest fluid state, the background energy E_b is:

$$\frac{dE_b}{dt} = P_{dyn} + F_{surf}^{diff} + F_{surf}^{adv} \tag{3.140}$$

where F_{surf}^{diff} is the diffusive mass flux and F_{surf}^{adv} is the advective mass flux across the surface S. These fluxes are provided by Winters et al. (1995).

Similarly to Eq. (3.135), the production of the background potential energy P_{dyn} becomes:

$$P_{dyn} = \frac{g}{V} \int_V F_{dyn} \, dV \tag{3.141}$$

When fluid is at rest, $|\nabla \rho| = -d\rho_{mol}/dz$ and $P_{dyn} \rightarrow P_{mol}$.

The Lorentz's (1955) idea was first applied by Thorpe (1977) for the one-dimensional density profile in lakes. The method was based on the sorting out of the elementary fluid volumes contributing to the fluid column by order of increasing density when the fluid has overturned. Thorpe recorded the distance that each particle has to be displaced by within the fluid column to reach its position in the stable profile. The root mean square value of these distances is now known as the Thorpe scale.

The extension of the method of sorting of density profile $\rho(z)$ for a three-dimensional fluid value was proposed by Winters et al. (1995), Bouruet-Aubertot et al. (2001) and Tseng and Ferziger (2001).

The adiabatic transformation which provides the stable density profile $\rho(z)$ is closely related to the available potential energy (see discussion above). According to Lorentz (1955), the available potential energy (APE) becomes:

$$E_a = E_p - E_b \tag{3.142}$$

where E_p is the potential energy per unit volume of the fluid before the transformation and E_b is the background potential energy. It is clear that the state of minimum potential energy is given by zero value of energy E_a, i.e. $E_a = 0$. For uniform stratification from Eq. (3.122) we have $E_a \sim 0.5\,g^2\rho/N^2$. The evolution equation for energy E_a results from the expressions for E_p and E_b (Staquet 2004). This equation shows that the dissipation rate of available potential energy E_a is the turbulent mixing rate $(P_{\text{dyn}} - P_{\text{mol}})$. It means that the determination of mixing requires the decomposition of the total potential energy E_p into a reversible part (available potential energy), which can be converted back to kinetic energy, and an irreversible part E_b (background potential energy), which increases through mixing.

References

Abramowitz M, Stegun IA (1975) Handbook of mathematical functions with formulas, graphs, and mathematical tables. Dover Publications, New York, 1045 p

Bouruet-Aubertot P, Koudella C, Staquet C, Winters KB (2001) Particle dispersion and mixing by breaking internal gravity waves: two-dimensional numerical experiments. Dyn Atmos Oceans 33(2):95–134

Carter GS et al (2008) Energetics of M_2 barotropic-to-baroclinic tidal conversion at the Hawaiian Islands. J Phys Oceanogr 38:2205–2223

Cushman-Roisin B (1994) Introduction to geophysical fluid dynamics. Prentice Hall, Englewood Cliffs, 320 p

Davies AM, Xing J (2007) On the influence of stratification and tidal forcing upon mixing in sill regions. Ocean Dyn 57:431–451

Golenko NN, Beszczyńska-Möller A, Piechura J, Walczowski W (1999) Some results of research on internal waves in the Stolpe Sill area. Oceanologia 41(4):537–551

Grimshaw R, Pelinovsky E, Talipova T (2003) Damping to large amplitude solitary waves. Wave Motion 37:351–364

Grimshaw R, Pelinovsky E, Talipova T, Kurkin A (2004) Simulation of the transformation of internal solitary waves on oceanic shelves. J Phys Oceanogr 34(12):2774–2791

Grimshaw R, Pelinovsky E, Talipova T (2007) Modelling internal solitary waves in the coastal ocean. Surv Geophys 28:273–298

Holliday D, McIntyre ME (1981) On potential energy density in an incompressible, stratified fluid. J Fluid Mech 107:221–225

Holloway PE, Pelinovsky EN, Talipova T, Barnes B (1999) A generalized Korteweg-de Vries model of internal tide transformation in the coastal zone. J Geophys Res 104:18333–18350

Ivey GN, Nokes RI (1989) Vertical mixing due to the breaking of critical internal waves on sloping boundaries. J Fluid Mech 204:479–500

Kang D, Fringer OB (2010) On the calculation of available potential energy in internal wave fields. J Phys Oceanogr 40:2539–2545

Kantorovitch LV, Krylov VI (1962) Approximate methods of advanced analysis. Gos Izd Fiz-Math Lit, Leningrad, 560 p (in Russian)

Klymak JM, Pinkel R, Liu CT, David L (2006) Prototypical solitons in the South China Sea. Geophys Res Lett 33:L11607

Krauss W (1966) Interne Wellen. Gebruder Borntraeger, Berlin, 248 p

Kundu PK, Cohen IM (2006) Fluids mechanics. Elsevier, Amsterdam, 759 p

Kurkina O, Talipova TG, Pelinovsky EN, Soomere T (2011) Mapping the internal wave field in the Baltic Sea in the context of sediment transport in shallow water. J Coast Res, Spec Issue 64:2042–2047

Lamb KG, Nguyen VT (2009) Calculating energy flux in internal solitary waves with an application to reflectance. J Phys Oceanogr 39:559–580

Lien RC, Henyey F, Ma B (2014) Large-amplitude internal solitary waves observed in the Northern South China Sea: properties and energetics. J Phys Oceanogr 44:1095–1115

Lorentz EN (1955) Available potential energy and the maintenance of the general circulation. Tellus 7:157–167

Massel SR (2013) Ocean surface waves; their physics and prediction, 2nd edn. World Scientific Publ, Singapore, 676 p

Moum JN, Klymak JM, Nash JD, Perlin A, Smyth WD (2007) Energy transport by nonlinear internal waves. J Phys Oceanogr 37:1968–1988

Munk W, Wunsch C (1998) Abyssal recipes II: energetics of tidal and wind mixing. Deep-Sea Res I 45:1977–2010

Pelinovsky EN, Talipowa TG, Stepanyants JA (1994) Modelling of nonlinear wave propagation in the horizontally inhomogeneous ocean. Fizika Atm i Okeana 30(1):79–83 (in Russian)

Pelinovsky E, Polukhina O, Talipova T (2007) Internal solitary waves. In: Grimshaw R (ed) Solitary waves in fluids, Chapter 4. WIT Press, Southampton, pp 85–110

Roberts J (1975) Internal gravity waves in the ocean. Marcel Dekker Inc, New York, 274 p

Staquet C, Sommeria J (2002) Internal gravity waves: from instabilities to turbulence. Annu Rev Fluid Mech 34:559–593

Staquet C (2004) Gravity and inertia-gravity internal waves: breaking processes and induced mixing. Surv Geophys 25:281–314

Stigebrandt A (1979) Observational evidence for vertical diffusion driven by internal waves of tidal origin in the Oslo Fjord. J Phys Oceanogr 9:435–441

Talipova TG, Pelinovsky EN, Kouts T (1998) Kinematic characteristics of an internal wave field in the Gotland Deep in the Baltic Sea. Okeanologiya 38(1):33–42 (in Russian)

Thorpe SA (1968) On the shape of progressive internal waves. Philos T R Soc A 263:563–614

Thorpe SA (1977) Turbulence and mixing in a Scottish Loch. Philos T R Soc A 286:125–181

Thorpe SA (2005) The turbulent ocean. Cambridge University Press, Cambridge, 484 p

Tseng YH, Ferziger JH (2001) Mixing and available potential energy in stratified flows. Phys Fluids 33(2):1281–1293

van Haren H (2012) The ocean's internal motion: a short overview of NIOZ thermistor string observations. J Sea Res 74:8–15

Winters KB, D'Asaro EA (1996) Two-dimensional instability of finite amplitude internal gravity wave packets near a critical level. J Geophys Res 94:12709–12719

Winters KB, Lombard PN, Riley JJ, D'Asaro EA (1995) Available potential energy and mixing in density-stratified fluids. J Fluid Mech 289:115–128

Woods JD (1968) Wave induced shear instability in the summer thermocline. J Fluid Mech 32(4):791–800

Chapter 4
Waves in a Continuously Stratified Sea of Varying Depth

4.1 Introduction

In the previous chapter, the propagation of internal waves was modelled under the assumption of constant water depth. However, the sea bottom only very seldom is horizontal. Therefore, varying of water depth and horizontal inhomogenities of the density field should be included for the prediction of internal waves propagation in many real situations, especially in shallow water. Observations of internal waves in the Andaman Sea (Osborne and Burch 1980), Sulu Sea (Apel et al. 1985), Massachusetts Bay (Halpern 1971) and Australian North West Shelf (Holloway et al. 1997) have shown that shoaling effects and local bottom changes may essentially influence internal waves evolution.

The available theoretical and numerical methods for prediction of the internal waves evolution for varying bottom topography and varying density distribution are summarized in Sect. 4.2. In Sect. 4.3, an important case of the internal waves propagation on the coastal slope is considered. Kinematics and dynamics of wave trains are modelled for short and long waves. Section 4.4 is devoted to the classical ray method for prediction of internal waves pattern when they approach the coastal zone at an arbitrary angle. Evolution of internal waves over large distances, when bathymetry and density structure are changing in space, is described in Sect. 4.5. The concept of conservation of energy flux is used to calculate the wave amplitude variation.

4.2 Review of Available Theoretical and Numerical Prediction Methods

There is a substantial literature on the various aspects of internal waves in a varying depth environment. Sandström (1969) presented the ray theory argument to construct a solution to the variable depth wave guide. It consists of waves in which the ampli-

S.R. Massel, *Internal Gravity Waves in the Shallow Seas*,
GeoPlanet: Earth and Planetary Sciences, DOI 10.1007/978-3-319-18908-6_4

tude and phase may vary in space. Larsen (1969) and Robinson (1969) developed the solution for the internal waves incident upon a knife edge and vertical barrier, respectively, which satisfies the radiation criteria for all modes transmitted through and reflected by the barrier. In Melnikov's (1982) paper, the stream function under the Boussinesq approximation was used for study of wave propagation on various depth forms, assuming that waves are stationary.

Another solution for the stationary internal waves propagating over the underwater sill of the cosinusoidal form was reported by Babij (1983). He found that the wave amplitudes in the vicinity of the sill are monotonically increasing functions of the sill height and the amplification of the amplitude depends on the ratio of sill width and incident wavelength.

Theoretical solution for the scattering of internal waves at finite topography was proposed by Müller and Liu (2000a, b). They used a mapping function based on ray tracing. The radiation conditions are satisfied and the energy is conserved. The fraction of the incoming energy flux transmitted onto the shelf and reflected back to the deep ocean and the distribution of these fluxes for particular modes are calculated. Moreover, a random superposition of incident waves is considered to model the scattering of random waves over topography. In mode number space, both the reflected and transmitted flux spectra are flatter than the incident spectrum, indicating a transfer from low to high mode numbers. However, the solution is restricted to the two-dimensional area and the stationary motion.

A special attention was given to the propagation of long internal waves. Observations made from space have demonstrated the influence of a variable bottom topography on the evolution of internal waves. The theoretical description of weakly nonlinear solitary waves in shallow water with continuous stratification of density is based on the Korteweg-de Vries (KdV) and extended Korteweg-de Vries (eKdV) equations. Because the oceanic stratification and water depth usually vary slowly in space and in time, this variability results in the variable-coefficient Korteweg-de Vries equation. For example, Pelinovsky and Shavratsky (1976) considered evolution of periodic and solitary internal waves of the finite amplitude for slowly changing water depth and stratification. They give a relationship between amplitude of the isopycnal displacement and water depth and Väisälä-Brunt frequency. Some detail results are shown for exponential and multilayered structure of this frequency.

Helfrich and Melville (1986) studied experimentally and theoretically the propagation and stability of long nonlinear internal waves over slope and shelf. A generalised Korteweg-de Vries equation, including the effects of nonlinearity, dispersion, dissipation was used in the calculations. A special attention has been given to so called turning point, when the coefficient α of the quadratic nonlinear term in the Korteweg-de Vries equation changes signs and the higher-order cubic nonlinearities become important.

Smyth and Holloway (1988), and Holloway et al. (1997, 1999) applied the extended Korteweg-de Vries (eKdV) equation to examine the transformation of internal tide on the Australian North West Shelf. It was found that the flow is dominated by nonlinearities and in the neighborhood of the observed shocks, the motion can be predicted by the hydraulic theory. The same equation was used by Pelinovsky et al.

(1994). They provided a simple derivation of the generalised Korteweg-de Vries (KdV) equation and developed a numerical scheme for integration of this equation. The model showed development of shocks and solitons as wave propagates shoreward over the continental slope and shelf. Also, the friction was found to be important in limiting the amplitudes of the evolving waves. Some examples are given for the shallow and deep regions of the Black Sea.

Talipova et al. (1998) discussed the dependence of the nonlinearity and dispersion parameters of the Korteweg-de Vries equation on the ocean bathymetry and vertical density structure. They showed that for Gotland Deep in the Baltic Sea the phase speed C and dispersion coefficient, β, depend mainly on water depth, while the nonlinearity coefficient, α, depends strongly on the vertical structure of Väisälä-Brunt frequency.

Grimshaw et al. (2004) showed that internal tides (or long-scale disturbances) transform into large-amplitude short-scale pulses of solitary-like forms. They applied the numerical simulation of the variable-coefficient extended Korteweg-de Vries equation for several oceanic shelves and demonstrated that the solitary waves may maintain its soliton-like form for large distances, which confirms solitons widely observed solitons in the oceans.

Kurkina et al. (2011) examined the geographical and seasonal distributions of kinematic parameters of long internal waves based on the Generalised Digital Environmental Model.

One of the very important parameters is the near-bottom velocity induced by internal waves for cases when the density jump layers are located in the vicinity of the seabed. Then re-suspension and erosion processes are very likely (see for example Bogucki and Redekopp 1999, 2008).

4.3 Waves Propagating on a Coastal Slope

The evolution and stability of the internal waves propagating and breaking over variable topography plays an important role for the coastal environment. Wave breaking is an effective mechanism for mixing nutrient-rich waters from the bottom with the biologically active upper layer (Helfrich and Melville 1986). Indirect observations suggest that wave energy, even originally tidal, could be transferred to the scale of turbulence by breaking. Within the relatively quiet interior, internal waves are generally assumed to be the main mixing agent.

Phillips (1966) was the first who considered the reflection of internal waves from the sloping sea bed. The anisotropic nature of internal waves, manifested by the fact that their direction of propagation depends on the frequency, results in the complex geometry of wave packet depending on the relationship of the incident wave angle to the slope inclination. Most efforts to describe the nonlinear evolution of the internal waves approaching the shore have employed Korteweg-de Vries (KdV) theory.

However, prior to discussing this complex problem, let us define the basic geometry of the internal waves incident and reflected from the bottom sloping upward at an angle β, under the condition of the buoyancy frequency N_0 independent of z. We assume that the incident wave and reflected wave are in the plane of the slope.

The incident vertical wave is given by:

$$w = W_i \exp[j(k_{x,i}x + k_{z,i}z - \omega_i t)] \tag{4.1}$$

where the subscript i refers to the incident wave field.

For further convenience, we represent the incident and reflected waves in terms of the stream function ψ. Thus we have:

$$\psi_i = \Psi_i \exp[j(k_{x,i}x + k_{z,i}z - \omega_i t)] \tag{4.2}$$

and

$$\psi_r = \Psi_r \exp[j(k_{x,r}x + k_{z,r}z - \omega_r t)] \tag{4.3}$$

On the bottom boundary, where $z = x \tan\beta$, the total stream function must be constant, i.e.:

$$\Psi_i \exp[j(k_{x,i}x + k_{z,i}z - \omega_i t)] + \Psi_r \exp[j(k_{x,r}x + k_{z,i}z - \omega_r t)] = 0 \tag{4.4}$$

In the expression above, the constant was assumed equal zero. To satisfy this expression, the following conditions should be true for all x along the bottom:

$$\left.\begin{aligned} &\omega_r = \omega_i \\ &k_{x,r} + k_{z,r}\tan\beta = k_{x,i} + k_{z,i}\tan\beta \end{aligned}\right\} \tag{4.5}$$

Thus, the frequency and component of the wave vector parallel to the bottom are both conserved under reflection (Pedlosky 2003). The angle of the reflected wave vector to the horizontal plane must have the same magnitude as for the incident wave vector.

Let the incident wave vector equal K_i. Then the components of the incident wave vector and the reflected wave vector along the bottom slope are $K_i\cos(\theta_i - \beta)$ and $K_r\cos(\theta_r + \beta)$, respectively (see Fig. 4.1). From the condition of no flow through the sea bottom it follows that:

$$K_i\cos(\theta - \beta) = K_r\cos(\theta + \beta) \tag{4.6}$$

when $\theta_i = \theta_r = \theta$, due to preservation of angles under reflection.

Following Pedlosky (2003) we define angle ϕ as:

$$\phi = \frac{\pi}{2} - \theta \tag{4.7}$$

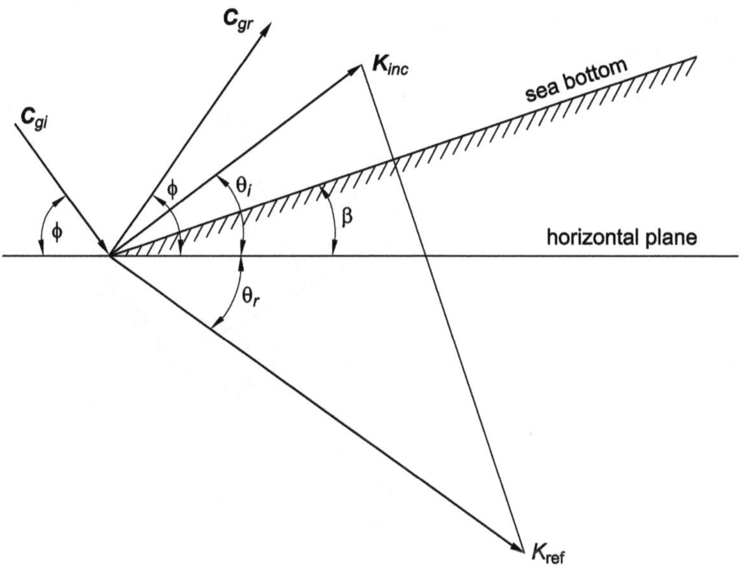

Fig. 4.1 Wavenumbers and group velocity vectors of incident and reflected waves when $\phi > \beta$

in which ϕ is the angle of the group velocity of the incident wave vector to the horizontal plane as well as the reflected wave vector (see Fig. 4.1). Thus from Eq. (1.53) we have:

$$\sin \phi = \frac{\omega}{N_0} \tag{4.8}$$

Substituting (4.7) into (4.6) we obtain:

$$\frac{K_r}{K_i} = \frac{\sin(\phi + \beta)}{\sin(\phi - \beta)} \tag{4.9}$$

where $\phi > \beta$. When angle ϕ approaches the angle β, the magnitude of the reflected wavenumber becomes infinite and the frequency is decreasing. Therefore, β can be considered a critical angle for the incident waves.

It should be noted that if the waves enter the wedge when $\phi > \beta$, the subsequent reflections from the bottom slope and the top surface will be forward. Waves advancing towards the apex of the wedge become shorter at every bottom reflection, and finally they dissipate in the apex of the wedge. We will discuss this case below in more detail.

Now let us consider the case when angle ϕ is smaller than β. In this case the right-hand side of Eq. (4.9) becomes negative. However, left-hand side of this equation should be always positive due to physical reasons. As we see in Fig. 4.2, in the case ($\phi < \beta$), the reflection must be backward, as opposite to the previous case ($\phi > \beta$) where the reflection is forward along the slope.

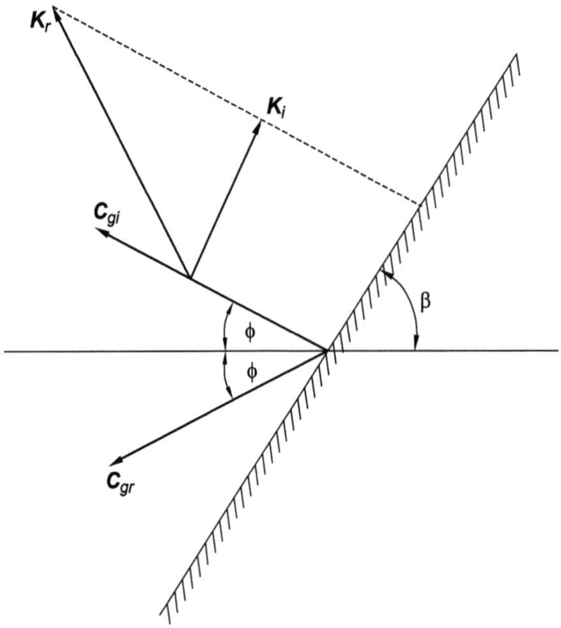

Fig. 4.2 Wavenumbers and group velocity vectors of incident and reflected waves from a sloping surface when $\phi < \beta$

The back reflection results in the following relation:

$$\frac{K_r}{K_i} = \frac{\sin(\beta + \phi)}{\sin(\beta - \phi)} \tag{4.10}$$

Let us now consider periodic internal wave of frequency ω propagating in an ocean of constant frequency N where the bottom slope is linear. Two rigid boundaries defined by $z = 0$ and $z = -\beta z$ for $x > 0$ formulate two-dimensional wedge geometry. Following Wunsch (1969), we represent the stream function as follows:

$$\psi(x, z, t) = \Psi(x, z)\exp(-i\omega t) \tag{4.11}$$

$$\frac{\partial^2 \Psi}{\partial z^2} - \frac{1}{c^2}\frac{\partial^2 \Psi}{\partial x^2} = 0 \tag{4.12}$$

where:

$$c^2 = \frac{\omega^2 - f^2}{N^2 - \omega^2} \tag{4.13}$$

Stream function ψ should satisfy the boundary conditions $\psi = 0$ at the sea surface and bottom surface. We also assume that frequency $N =$ constant.

For wave propagating up-slope, when the wedge angle β is smaller than the critical angle $\alpha = \arctan(1/c)$ (subcritical case), the stream function amplitude becomes:

$$\Psi(x, z) = 2A \sin\left[\frac{1}{2} q \ln\left(\frac{cx - z}{cx + z}\right)\right] \exp\left[-\frac{i}{2} q \ln(c^2x^2 - z^2)\right] \quad (4.14)$$

in which:

$$q = \frac{2n\pi}{\ln \Delta}, \qquad \Delta = \frac{c + \beta}{c - \beta} \quad (4.15)$$

Note that singularities at $x = z = 0$, as well as along the line $z = -cx$ appear.

When $\beta/c \ll 1$ (shallow wedge) and $cx \gg z$, Eq. (4.13) can be simplified as follows:

$$\Psi \approx 2iA \exp\left[\frac{in\pi c}{\beta} \ln(cx)\right] \sin\left(\frac{n\pi z}{\beta x}\right) \quad (4.16)$$

The local horizontal wavenumber $k_x = n\pi c/\beta x$ and amplitudes of both velocity components are:

$$u = -2iA \left\{\frac{n\pi}{\beta x} \exp\left[-\frac{in\pi c}{\beta} \ln(cx)\right] \cos\left(\frac{n\pi}{\beta}\frac{z}{x}\right)\right\} \quad (4.17)$$

and

$$w = -2iA \exp\left[-\frac{in\pi c}{\beta} \ln(cx)\right] \left[\frac{n\pi}{\beta x} \sin\left(\frac{n\pi}{\beta}\frac{z}{x}\right) + \frac{n\pi z}{\beta x^2} \cos\left(\frac{n\pi}{\beta}\frac{z}{x}\right)\right] \quad (4.18)$$

As the corner is approached the wavelength decreases and the amplitudes of velocities increase. The structure of these waves is similar to the standing modes (Wunsch 1968, 1969).

The reflection of two-dimensional wave train of finite amplitude from an inclined plane boundary has been discussed by Thorpe (1987), specifically when $\phi > \beta$. Initial wave packet was chosen as exact solutions of the incident and reflected first-order waves. The interaction of first-order waves generates additional components. In fact, strong nonlinearity was found at adjacent values of ϕ and β. The effect of nonlinearity results in reducing the steepness of the incident wave.

The problem of wave reflection from an uniform slope at second order was investigated further by Thorpe (1997). He found that the resonant interactions cannot occur, unless the slope inclination β is smaller than the inclination of the wave group velocity vector ϕ. The interaction generates a nonzero Eulerian up slope flow, and reflected waves generate a Stokes drift parallel to the isobaths.

In coastal region, the internal waves are usually considered nonlinear long waves. Therefore, the Korteweg-de Vries (KdV) theory is used to describe the evolution of these waves is shallow waters as well as propagating over coastal slopes. Helfrich and Melville (1986) published results of theoretical and experimental studies on long internal waves propagating over slope-shelf topography. A generalised Korteweg-de Vries (KdV) equation, including the nonlinearity, dispersion, dissipation and varying topography was solved numerically for single and rank-ordered pairs solitary waves. Very good agreement of the theoretical results with laboratory experiments in a salt-stratified wave tank was shown. Good results were obtained even for a single solitary wave incident on slope-shelf topography of large dimension when the topographic effects dominate nonlinearity and dispersion. Moreover, comparisons of the theoretical model with the experiments when strong breaking occurs, showed surprisingly good agreement.

The authors also presented some conclusions on the application of developed theoretical model for real oceanographic situations. However, they are not sure to which extent a laminar damping in the laboratory tanks properly reflects turbulent eddy viscosity in the ocean.

Laboratory experiments and field observations were used by Boegman et al. (2005) to examine the energy flux in a two-layer stratified system subjected to a single forcing event. The energy loss along the slope, the mixing efficiency were modelled using appropriate forms of an internal Iribarren number which is defined as the ratio of the boundary slope to the wave slope. Therefore, for classifying the breaker type we have:

$$\xi_b = \frac{\tan\beta}{\sqrt{\dfrac{H_b}{L_\infty}}} \quad \text{or} \quad \xi_\infty = \frac{\tan\beta}{\sqrt{\dfrac{H_\infty}{L_\infty}}} \tag{4.19}$$

in which $\tan\beta$ is the bottom slope, H_b and H_∞ are the wave height at the breaker point and off-shore, respectively. The experiments showed that plunging breaker type was dominating.

Authors also made an attempt to represent the dynamics of internal wave breaking in terms of readily measured variables. In particular, for solitary internal wave, the following form of the Iribarren number ξ may be useful for classification of the breaker type:

$$\xi = \frac{\tan\beta}{\sqrt{\dfrac{\alpha\,\eta_0}{C_0}}} \tag{4.20}$$

where α is the nonlinear coefficient, C_0 is the linear phase velocity and η_0 is the maximum excursion of the internal layer interface.

It was shown experimentally that a single wave breaking event resulted in from 10 to 75 % of the incident wave energy lost due to dissipation and mixing while remaining part of energy was reflected. However, the level of the energy loss was strongly dependent on the breaker type. The same conclusion is valid for efficiency of mixing.

The breaking internal waves induce the mixing which is significant for the overall dynamics of coastal zone. In particular, Eriksen (1985), Ivey and Nokes (1989), Garrett and Gilbert (1988) discussed the mixing that can occur when internal waves are reflected off the sloping bottom. They focused on the case of the waves of critical frequency when the slope of the group velocity vector of the incident waves is equal to the bottom slope β. They found that the turbulent mixing is confined to the bottom boundary layer, and the mixing efficiency had a maximum value of 0.20. The mixing efficiency is defined as the ratio of the increase in potential energy due to mixing to the loss of the kinetic energy by the incident waves. For a more detailed description of the mixing processes on the sloping bottom I refer the reader to the book by Thorpe (2005).

4.4 Ray Method for Internal Waves

In the section above, a transformation of internal waves on the sloping bottom was considered mostly for the two-dimensional case. Similarly to the surface waves, in general case when internal waves are approaching the coast line under arbitrary angle, the classical ray methods based on the geometrical optics principles are useful. So let us assume that small-amplitude waves propagate in the non-uniform, but slowly varying medium. Any local property of the wave train, i.e. wave velocity, wave pressure, isopycnal displacement and other, can be specified by the following function:

$$\eta(\mathbf{x}, t) = A(\mathbf{x}, t) \exp[i\varphi(\mathbf{x}, t)] \tag{4.21}$$

where $A(\mathbf{x}, t)$ varies in space and time slowly, compared with the wavelength and period of waves. The $\varphi(\mathbf{x}, t)$ is a rapidly varying oscillatory phase function. The propagation of points of constant phase is given by $\varphi(\mathbf{x}, t) = \text{const}$.

Let us now define the local wavenumber $\mathbf{k}$ and frequency ω by (Massel 1989):

$$\mathbf{k} = \nabla\varphi \quad \text{and} \quad \omega = -\frac{\partial\varphi}{\partial t} \tag{4.22}$$

in which ∇ is the usual gradient operator with respect to the components of $\mathbf{x}$:

$$\nabla = \frac{\partial}{\partial x}\mathbf{i} + \frac{\partial}{\partial y}\mathbf{j} + \frac{\partial}{\partial z}\mathbf{k} \tag{4.23}$$

From the first expression in (4.22) follows immediately that $\nabla \times \mathbf{k} = 0$. Thus, the local wavenumber vector is irrotational. After eliminating the phase φ we obtain:

$$\frac{\partial \mathbf{k}}{\partial t} + (\nabla_{\mathbf{k}}\omega) \cdot (\nabla \mathbf{k}) = -\nabla\omega \tag{4.24}$$

The symbol $\nabla_{\mathbf{k}}$ denotes derivatives taken with respect to the components of $\mathbf{k}$, whereas ∇ is the usual gradient operator taken as derivative with respect to the components of $\mathbf{x}$.

If we restrict ourselves for a moment to one dimension, we get:

$$\frac{\partial k_x}{\partial t} + \frac{\partial \omega}{\partial x} = 0 \tag{4.25}$$

In previous chapters it has been shown that the intrinsic frequency ω and the wavenumber $\mathbf{k}$ are related by the dispersion relation, which depends not only on the local wavenumber, but also on the local wave depth, current velocity and water density. If the medium itself is moving with velocity $\mathbf{U}$, the frequency of waves passing a fixed point is:

$$\omega = \Omega(\mathbf{k}, \mathbf{x}) + \mathbf{k} \cdot \mathbf{U} \tag{4.26}$$

in which the extrinsic frequency $\Omega = \Omega(\mathbf{k}, \mathbf{x})$ satisfies the dispersion relation for waves in a stationary ambient.

Therefore, from Eq. (4.24) we have:

$$\frac{\partial k_i}{\partial t} + Cg_j(\mathbf{x}, \mathbf{k}, t)\frac{\partial k_j}{\partial x_i} = -\frac{\partial \Omega}{\partial x_i}, \quad i = 1, 2, 3 \tag{4.27}$$

in which:

$$Cg_j = \frac{\partial \Omega}{\partial k_j} \tag{4.28}$$

Equation (4.27) can be rewritten in the canonical form as follows:

$$\frac{dk_i}{dt} = -\frac{\partial \Omega}{\partial x_j} \quad \text{and} \quad \frac{dx_i}{dt} = \frac{\partial \Omega}{\partial k_i} \tag{4.29}$$

The equations above are canonical equations for the rays. Moreover, it should be noted that Eq. (4.29) are identical with Hamilton equations in mechanics. The vectors $\mathbf{x}$ should be interpreted as coordinates and vectors $\mathbf{k}$ represent the momenta. Then, the frequency $\Omega(\mathbf{k}, \mathbf{x}, t)$ becomes the Hamiltonian of the system (Whitham 1974).

Following Sutherland (2010), we consider now the propagation of interfacial waves in a two-layer fluid. We suppose that waves move on a constant slope, approaching the elongated beach. Let the ray form oblique angle versus normal direction to the beach. The upper layer of fluid has a constant depth h_1 and density ρ_1, far from the shore. The lower layer has a density ρ_2 and the depth $h_2(x)$ decreases as:

$$h_2(x) = \beta|x| - h_1 \tag{4.30}$$

in which β is the bottom slope. Thus, interface intersects with the bottom at $x = -h_1/\beta$.

As the topography is independent of y and motion is stationary, the frequency $\omega = \omega_0 = \text{const}$ and component of wavenumber $k_y = k_{y0} = \text{const}$. However, wavenumber component k_x is changing as the water depth is decreasing. The corresponding dispersion relation can be found by a generalisation of the dispersion relation (2.8), i.e.:

$$\omega^2 = \frac{g|k_h|(\rho_2 - \rho_1)}{\rho_1 \coth[|k_h|h_1] + \rho_2 \coth[|k_h|h_2]} \tag{4.31}$$

where $k_h^2 = k_x^2 + k_{y0}^2$. The wavenumber k_x should change as water depth h_2 varies in a such a way that frequency ω remains constant with the constant value of wavenumber k_{y0} and densities ρ_1 and ρ_2.

The trajectory of wave rays is following from Eq. (4.29), i.e.:

$$\frac{dx}{dt} = \frac{\partial \omega}{\partial k_x} = C_{gx} \quad \text{and} \quad \frac{dy}{dt} = \frac{\partial \omega}{\partial k_y} = C_{gy} \tag{4.32}$$

or

$$\frac{dy}{dx} = \frac{\partial \omega / \partial k_y}{\partial \omega / \partial k_x} \tag{4.33}$$

Thus, the path of waves is now independent of their position in time. As the right-hand side of Eq. (4.33) does not depend on y, the solution of this equation becomes:

$$y(x) = y_0 + \int_{x_0}^{x} \frac{\partial \omega}{\partial k_y} \left(\frac{\partial \omega}{\partial k_x} \right)^{-1} dx \tag{4.34}$$

Usually, the internal waves close to the beach are long waves when $|k_h|h_1 \ll 1$ and $|k_h|h_2 \ll 1$. Thus, from the generalisation of Eq. (2.8) we have:

$$\omega = \sqrt{\frac{g(\rho_2 - \rho_1)h_1 h_2}{\rho_1 h_2 + \rho_2 h_1}}\, |k_h| \tag{4.35}$$

Now, we go back to the internal waves moving along x-direction, perpendicularly to the coast, in vertically varying background flow $U(z)$ and buoyancy frequency $N(z)$. This is the similar case as discussed in section above. Now the wave numbers are constant, and $k_y = 0$ and $k_x = k_{x0}$. For this condition, from Eq. (4.26) we have $\omega = \omega_0 + k_{x0} U$. The wave path in the $(x_1 = x, x_3 = z)$ space is determined by two relationships resulting from Eq. (4.29):

$$\frac{dx}{dt} = C_{gx} + U \quad \text{and} \quad \frac{dz}{dt} = C_{gz} \tag{4.36}$$

or

$$\frac{dz}{dx} = \frac{C_{gz}}{C_{gx} + U} \tag{4.37}$$

in which components of the group velocity are given by Eq. (1.60). If $U = 0$, the path of internal waves is given as:

$$\frac{dz}{dx} = \frac{C_{gz}}{C_{gx}} = -\frac{k_x}{k_z} \tag{4.38}$$

From Eq. (1.52) it follows that when observed frequency $\omega \to N_0$, the wavenumber component $k_z \to 0$ and vertical component of the group velocity $C_{gz} \to 0$.

Dynamics of the internal waves approaching the coast can be discussed, similarly to the surface waves, by applying the conservation laws (Massel 1989). One of the basic conservation laws is the conservation of wave action. This principle is generally applied for any non-dissipative disturbances in a moving medium. If the mean current velocity U is not uniform, the intrinsic frequency σ and the wavenumber k will vary in space and time. Thus, although the wave action $\mathscr{A} = E/\sigma$ is conserved, the average energy per unit mass E is not. For example, for surface waves the local rate of change of wave action is balanced by the convergence of the flux of action as follows (Massel 1989):

$$\frac{\partial}{\partial t}\left(\frac{E}{\sigma}\right) + \nabla_h \left[\frac{E}{\sigma}(\mathbf{U} + \mathbf{C_g})\right] = 0 \tag{4.39}$$

where $\nabla_h = \frac{\partial}{\partial x}\mathbf{i} + \frac{\partial}{\partial y}\mathbf{j}$.

For the small-amplitude waves propagated in x-direction we can write:

$$\mathscr{A} = \frac{\overline{E}}{\sigma} = \frac{\overline{E}}{\omega - k_x U} \tag{4.40}$$

From Eq. (3.117) it follows that $\overline{E} \approx (NA)^2$, where A is the isopycnal displacement amplitude. If we assume that internal waves do not spread substantially in the horizontal direction as they move (as for the case of horizontally periodic waves), Eq. (4.39) should be rewritten as (Sutherland 2010):

$$\frac{\partial(\mathscr{A})}{\partial t} = -\frac{\partial}{\partial z}(C_{gz}\mathscr{A}) \tag{4.41}$$

After substituting (4.40) into (4.41), for stationary current we obtain:

$$C_{gz}\frac{(NA)^2}{\sigma} = C_{gz}\frac{(NA)^2}{\omega - k_x U} = \text{constant} \tag{4.42}$$

The amplitude A increases as N or C_{gz} decreases. However, when the nonlinear effects become important and wave breaking appears, the ray theory predictions are not valid.

4.5 Nonlinear Internal Waves Propagation in a Horizontally Inhomogeneous Sea

When internal waves propagate on large distances in the ocean, their basic characteristics vary in time and in space. Due to changing of water depth and density structure, wave shape and wave amplitude slowly change too. In the linear approximation, the evolution of the monochromatic waves was considered by Keller and Mov (1969), and Miropolsky (1974). However, observations showed that the nonlinear effects play an important role in the evolution of internal wave and cannot be omitted. To get a simple estimation of variation of internal wave parameters in space, Pelinovsky and Shavratsky (1976) considered variation of the nonlinear and non-sinusoidal wave shapes, such as the cnoidal and solitary wave profiles. If the dissipation of wave energy is neglected, the mean energy flux F can be considered as a constant value. Thus we have:

$$F = \int_0^T \int_0^{h(x)} pu\, dz\, dt = \text{const} \tag{4.43}$$

After representing pressure p and velocity u for the cnoidal wave we obtain (Pelinovsky and Shavratsky 1976):

$$\frac{Y(m)}{C_0 q^6 \mathscr{L}} = \text{const} \tag{4.44}$$

in which:

$$\mathscr{L}^{-1} = \int_0^h \left(\frac{d\Phi}{dz}\right)^2 dz \tag{4.45}$$

$$q^3 = \frac{\int_0^h \bar{\rho}(z)\left(\frac{d\Phi}{dz}\right)^3 dz}{\int_0^h \bar{\rho}(z)\Phi^3 dz} \tag{4.46}$$

and

$$Y(m) = \mathbf{K}^4(m)\left[\frac{4-2m^2}{3}\frac{\mathbf{E}(m)}{\mathbf{K}(m)} - \frac{1-m^2}{3} - \left(\frac{\mathbf{E}(m)}{\mathbf{K}(m)}\right)^2\right] \tag{4.47}$$

where $\mathbf{K(m_0)}$ and $\mathbf{E(m_0)}$ are the complete elliptic integrals of the first and second kind, respectively (Abramowitz and Stegun 1975). Note that vertical axis in the integrals above is directed downwards.

Function $\Phi(z)$ describes vertical density structure of the fluid being solution of the linear boundary value problem (see also Chap. 3):

$$\frac{d^2\Phi(z)}{dz^2} + \frac{N^2(z)}{C_0^2}\Phi(z) = 0, \quad \Phi(0) = \Phi(h) = 0 \tag{4.48}$$

For some values of the parameter m, Eq. (4.47) can be considerably simplified. If $m \ll 1$, $Y(m) \sim m^4$ and cnoidal wave profile approaches to sinusoidal one. Then from Eq. (4.44) it follows that the wave amplitude, horizontal and vertical velocity amplitudes and variations of fluid density are:

$$A(x) \approx \sqrt{\frac{\mathscr{L}}{C_0^3}} \tag{4.49}$$

$$U(x) \approx \sqrt{\frac{\mathscr{L}}{C_0}}\left|\frac{d\Phi}{dz}\right|_{\max} \qquad W(x) \approx \sqrt{\frac{\mathscr{L}}{C_0^3}} \tag{4.50}$$

and

$$\rho(x) \approx \sqrt{\frac{\mathscr{L}}{C_0^3}}\left|\Phi\frac{d\bar{\rho}}{dz}\right|_{\max} \tag{4.51}$$

For another extreme case, when $m \to 1$, wave profile approaches the solitary wave for which we have:

$$\left.\begin{aligned} A(x) &\approx W(x) \approx q\mathscr{L}^{2/3}C_0^{-4/3} \\ U(x) &\approx q\mathscr{L}^{2/3}C_0^{-1/3}\left|\frac{d\Phi}{dz}\right|_{\max} \end{aligned}\right\} \tag{4.52}$$

The expressions above are valid for arbitrary density distribution.

In Chap. 3 it was shown that the simple exponential distribution of fluid density (see Eq. 3.1) appears to be a good approximation for some real distributions. For this distribution, the Väisälä-Brunt frequency $N(z) = N_0 = \text{const}$ and eigenfunction $\Phi(z) = \sin(n\pi z/h)$. Substituting this into Eqs. (4.45) and (4.46) we have:

$$\mathscr{L} = \frac{2h}{(n\pi)^2}, \quad q^2 \sim \begin{cases} \dfrac{gh^2}{N_0^2} & \text{for odd modes} \\[2ex] \dfrac{g^2h}{N_0^4} & \text{for even modes} \end{cases} \tag{4.53}$$

Therefore, for sinusoidal wave profile ($m \ll 1$), the wave amplitude A varies as follows:

$$A(x) \sim h^{-1}(x) N_0^{-3/2}(x) \tag{4.54}$$

When the water depth is diminishing, the amplitude of the internal wave is decreasing faster than the amplitude of the surface wave, i.e. $A \sim h^{-1/4}$ (Green's law). If the Väisälä-Brunt frequency N is going to be smaller, the water density becomes more uniform and the buoyancy force is decreasing, which finally results in a smaller wave amplitude.

On the other hand, if parameter $m \to 1$, the amplitude of soliton or solitary wave becomes:

$$\left.\begin{aligned} A(x) &\sim h^{-4/3}(x) N_0^{-2/3}(x) \text{ for odd modes} \\ A(x) &\sim h^{-1}(x) \qquad\qquad\quad \text{for even modes} \end{aligned}\right\} \tag{4.55}$$

The three-layer density distribution is frequently observed in nature in shallow waters such as Baltic Sea (Krauss 1966). So, let us consider the density distribution when the Väisälä-Brunt frequency $N(z) = N_0(x) \neq 0$ only within the layer $h_1(x) \leq z \leq h_1(x) + \Delta(x)$, and $N(z) = 0$ for $0 < z < h_1(x)$ and $h_1(x) + \Delta(x) \leq z \leq h(x)$, in which $\Delta(x)$ is the variation of the thickness of non-uniform layer. Also, we assume that $h_1(x), \Delta(x) \ll h(x)$.

For the case $m \ll 1$, from Eq. (4.49) we obtain:

$$A(x) \sim N_0^{-3/2}(x)\, \Delta^{-1}(x) \tag{4.56}$$

In contrast to Eq. (4.54) the wave amplitude now depends on the thickness of the pycnocline.

In another limiting case, i.e. $m \to 1$, for the three-layer density distribution, we have (Pelinovsky and Shavratsky 1976):

$$A(x) \sim (N_0(x)\Delta(x))^{-4/3} h^{-1/3}(x) \left\{ 1 - \left[1 + \left(\frac{\Delta(x)}{n\pi h(x)} \right)^2 \right]^{-3/2} \right\}^{1/3} \tag{4.57}$$

where n is the mode number.

It should be noted that for arbitrary thickness of the pycnocline Δ and depth h_1, variation of wave amplitude depends on the mode number n. For higher mode numbers, effective thickness of the pycnocline Δ/n is decreasing and for higher modes and for deep pycnocline we obtain:

$$A(x) \sim N_0^{-4/3}(x)\, h^{-1/3} (h_1 \Delta)^{-2/3} \tag{4.58}$$

If the pycnocline is located close to the sea surface, from Eq. (4.57) we obtain:

$$A(x) \sim N_0^{-4/3}(x)\, h^{-1/3}(x)\, \Delta^{-4/3}(x) \tag{4.59}$$

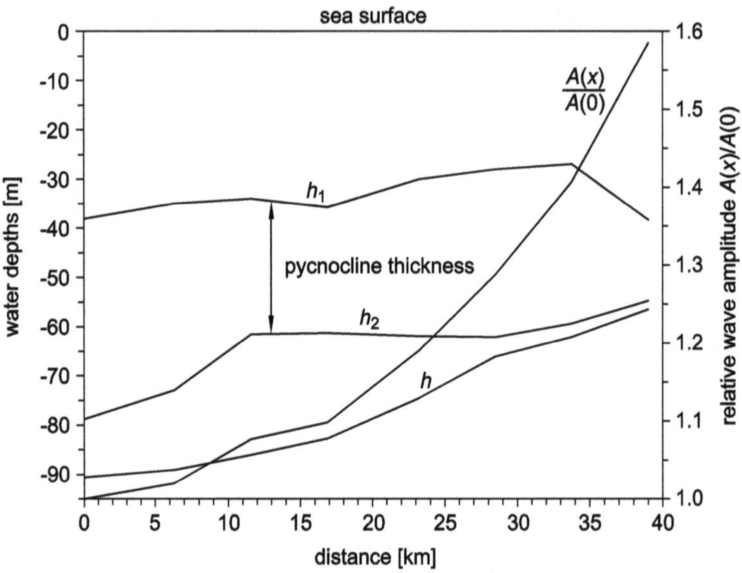

Fig. 4.3 Relative amplitude of a solitary wave as a function of distance for the eastern slope of Stolpe Sill (Southern Baltic)

It should be noted that the effect of variation of the pycnocline thickness is stronger than the water depth variation.

In Fig. 4.3, variation of the internal wave amplitude is illustrated for solitary internal wave propagating on the eastern slope of the Stolpe Sill in the Southern Baltic for data recorded in February 2004. The relative wave amplitude was calculated as:

$$A_r(x) = \frac{A(x)}{A(0)} \tag{4.60}$$

in which $A(0)$ is an amplitude at the tow of the slope, where initial water depth equals $h \approx 90\,\text{m}$. The water depth at the sill top is about 56 m, and the average bottom inclination equals about 0.00085. Under the assumption of the three-layer density structure, the Väisälä-Brunt frequency $N_0(x)$ varies from 0.03 to 0.05 rad s^{-1}. The function $A_r(x)$ in Fig. 4.3 was given for the first mode, $n = 1$.

In the discussion above, the shape of the internal wave profile was prescribed. We simply assumed the shape of cnoidal wave which in the extreme cases ($m \ll 1$ or $m \to 1$) takes very simple sinusoidal or solitary (soliton) form, respectively. However, when the wave profile is not prescribed, the horizontal non-homogenity of the water basin can be taken into account, assuming that the energy flux is conserved (see Eq. 4.43) along the wave rays and the wave reflection is negligible. If the Earth's rotation is omitted, generalised KdV equation for a horizontally unhomogeneous fluid becomes (Pelinovsky et al. 1994):

$$\frac{\partial \eta}{\partial t} + C_0 \frac{\partial \eta}{\partial s} + \alpha \eta \frac{\partial \eta}{\partial s} + \beta \frac{\partial^3 \eta}{\partial s^3} + \frac{C_0}{2} \frac{1}{Q} \frac{dQ}{ds} \eta = 0 \tag{4.61}$$

in which:

$$Q = \Delta_r M C_0^3, \quad M = \int_{-h}^{0} \left(\frac{d\Phi}{dz} \right)^2 dz \tag{4.62}$$

and Δ_r is the distance between the wave rays.

Let us define the new variables as follows:

$$\zeta = P\eta, \quad P = \sqrt{\frac{MC_0\Delta_r}{(MC_0\Delta_r)_0}} \tag{4.63}$$

where M_0, C_{00} and Δ_{r0} are the initial values of the parameters. After substituting in Eq. (4.61) we obtain:

$$C_0 \frac{\partial \zeta}{\partial l} + \frac{\alpha \zeta}{C_0 P} \frac{\partial \zeta}{\partial s} + \frac{\beta}{C_0^3} \frac{\partial^3 \zeta}{\partial s^3} = 0 \tag{4.64}$$

in which $s = \int dl/C_0 - t$ and l is the distance along the wave ray.

If $\hat{\alpha} = \alpha/PC_0^2$ and $\hat{\beta} = \beta/c_0^3$, Eq. (4.64) can be presented in a standard form of the KdV equation with changing coefficients:

$$\frac{\partial \zeta}{\partial l} + \hat{\alpha} \frac{\partial \zeta}{\partial s} + \hat{\beta} \frac{\partial^3 \zeta}{\partial s^3} = 0 \tag{4.65}$$

The coefficients $\hat{\alpha}$ and $\hat{\beta}$ depend on the fluid stratification, water depth and the curvature of the wave rays.

Pelinovsky et al. (1994) presented the numerical solution of the Eq. (4.65) for the shallow water in the southern Crimea shelf. Internal wave of period $T = 5$ min and initial amplitude $A = 1.7$ m propagated in a two-layer fluid with the density jump of $\Delta\rho/\rho \approx 4 \times 10^{-3}$ at water depth ≈15 m. Total water depth was about 60 m. Due to nonlinear effects, the wave profile has changed substantially even on the distance of 4–8 Nm. Solitons propagating with speed bigger than C_0 appear on the main wave crest.

On larger distances, solitons of higher amplitudes are catching up the solitons of lower amplitudes and the wave profile becomes very close to the sinusoidal one again. It is well known that evolution of a nonlinear wave train, in the absence of dissipation effects, exhibits the Fermi-Pasta-Ulam recurrence phenomenon, where the modulation periodically increases and decreases, and the wave form returns periodically to the previous form (Massel 2007). A similar behaviour of the internal wave profile resulted from the KdV equation is observed. However, for the extended KdV equation (4.61), the full recurrence of the wave form does not always happen.

When the water depth and water stratification vary slowly in the horizontal direction, the structure of the internal waves also varies slowly. However, to the first approximation, the dynamics of internal waves can be still calculated from the extended KdV equation when the Earth's rotation is omitted and all coefficients are functions of the horizontal distance:

$$\frac{\partial \eta}{\partial t} + \left(C_0 + \alpha\eta + \alpha_1\eta^2\right)\frac{\partial \eta}{\partial x} + \beta\frac{\partial^3 \eta}{\partial x^3} + \frac{C_0}{2Q}\frac{dQ}{dx}\eta = 0 \tag{4.66}$$

Grimshaw et al. (2004) employed this equation to simulate the dynamics of internal solitary waves taking into account real variability of the oceanic parameters. If the background environment varies slowly in comparison with an individual solitary wave, the propagating wave has a soliton-like form with varying amplitude and phase. The numerical simulations were applied for description of internal waves for the NWS of Australia, the Arctic shelf in the Laptev Sea and for the Malin shelf edge. It is interesting to note that the internal solitary wave has a soliton like form for a distance ranging from 80 to 140 km, depending on the simulation region. The corresponding lifetime of internal solitary waves may vary from 10 to 50 h.

For internal wave propagated the long distances over several inertial periods, the effect of the Earth's rotation is significant. The relevant equation in which the effects of slowly varying topography and Earth's rotation are combined is the so called rotated extended Korteweg-de Vries equation (reKdV). This equation, which is the variable coefficient Ostrovsky's equation, has the form (Ostrovsky 1978; Holloway et al. 1999; Grimshaw et al. 2014):

$$\frac{\partial}{\partial x}\left[\frac{\partial \eta}{\partial t} + \left(C_0 + \alpha\eta + \alpha_1\eta^2\right)\frac{\partial \eta}{\partial x} + \beta\frac{\partial^3 \eta}{\partial x^3} - \frac{C_0}{Q}\frac{dQ}{dx}\eta\right] = \frac{f^2}{2C_0}\eta \tag{4.67}$$

where:

$$Q = \sqrt{\frac{C_{00}^3}{C_0}\frac{\int\limits_{-h}^{0}\left(\frac{d\Phi_0}{dz}\right)^2 dz}{\int\limits_{-h}^{0}\left(\frac{d\Phi}{dz}\right)^2 dz}} \tag{4.68}$$

Values with subscript 0 are the values at some fixed point x_0. Holloway et al. (1999) used the Eq. (4.67) to simulate the nonlinear evolution of the initially sinusoidal internal tide for the North West Shelf (NWS) of Australia. Role of both quadratic and cubic nonlinearity, the Earth's rotation, and frictional dissipation were discussed. The cubic nonlinear effects as well as the Earth's rotation are important for proper description of the internal tide evolution over long distances. Predicted theoretical forms of the internal waves have been found to be consistent with observations of the internal tides on the NWS.

In the simulations of the internal waves in the South China Sea, a secondary wave packet trailing behind the leading internal wave was observed with the structure of the KdV-like modular bore. This is due to enhanced radiation from the leading wave (Grimshaw et al. 2014).

Internal waves propagating over variable topography on such long distances undergo substantial deformations which sometimes reach the breaking limit when the wave particle speed exceeds the wave propagation speed and a trapped core is formed within the internal solitary waves (Lien et al. 2014). Helfrich (1992) found experimentally, that internal wave breaks when the normalised maximum vertical displacement $\bar{\eta}_{\max}$ exceeds 0.4, independent of the bottom slope. The displacement is defined as follows:

$$\bar{\eta}_{\max} = \eta_{\max}(h - h_m) \tag{4.69}$$

where h is the water depth, and $(h - h_m)$ is the thickness of the lower layer when h_m is the initial depth of the isopycnal undergoing maximum vertical displacement.

At present, the South China Sea is known as a "hot spot" for observation of internal waves. These waves are formed by conversion from internal tides generated in the Luzon Strait. Satellite images showed a single depression in the deep part of the basin, and waves of multiple depressions on the Dongsha Plateau, between the continental slope and continental shelf (Lien et al. 2014). Observations in this region showed that the $\bar{\eta}_{\max}$ is between 0.3 and 0.4. According to Helfrich (1992) when $\bar{\eta}_{\max} < 0.3$, the internal waves are stable.

In contrast to these experimental observations, numerical model simulations by Vlasenko and Hutter (2002) suggest that the wave breaking is a function not only of the normalised maximum vertical displacement but also the bottom slope. They studied evolution and transformation of internal solitary waves propagating over the continental slope. Theoretical and numerical studies have been focused on the overturning and breaking of large amplitude waves. It was found that wave breaking occurs when $\bar{\eta}_{\max} > 0.4(h - h_m)$. Before breaking the horizontal orbital velocity exceeds the phase speed of the internal solitary waves and the top of the propagating bore outstrips the wave trough. The heavier and denser water penetrates into the relatively light water layers. For slope angles $0.52° < \beta < 21.8°$, Vlasenko and Hutter (2002) found that breaking event arises when nondimensional displacement $\bar{\eta}_{\max} \approx 0.8°/\beta + 0.4$. However, this condition has not been confirmed by Lien et al. (2014) observations in the South China Sea.

References

Abramowitz M, Stegun IA (1045) Handbook of mathematical functions with formulas, graphs, and mathematical tables. Dover Publications, New York

Apel JR, Holbrook JR, Liu AK, Tsai JJ (1985) The Sulu Sea internal soliton experiment. J Phys Oceanogr 15:1625–1651

Babij MB (1983) Transformation and generation of internal waves over the underwater sill in the stratified ocean. In: Theoretical studies of the wave processes in the ocean. Sevastopol, pp 101–105 (in Russian)
Boegman L, Ivey GN, Imberger J (2005) The degeneration of internal waves in lakes with sloping topography. Limnol Oceanogr 50(5):1620–1637
Bogucki DJ, Redekopp LG (1999) A mechanism for sediment resuspension by internal solitary waves. Geophys Res Lett 26(9):1317–1320
Bogucki DJ, Redekopp LG (2008) Climate of long internal waves and resuspension on the coastal shelf. Oceanologia 50(1):5–21
Eriksen CC (1985) Implications of ocean bottom reflection for internal wave spectra and mixing. J Phys Oceanogr 15:1145–1156
Garrett C, Gilbert D (1988) Estimates of vertical mixing by internal waves reflected off a sloping bottom. Small-scale turbulence and mixing in the ocean. In: Proceedings of the 19th international Liege colloquium on ocean hydrodynamics, pp 405–424
Grimshaw R, Pelinovsky E, Talipova T, Kurkin A (2004) Simulation of the transformation of internal solitary waves on oceanic shelves. J Phys Oceanogr 34(12):2774–2791
Grimshaw R, Guo C, Helfrich K, Vlasenko V (2014) Combined effect of rotation and topography on shoaling oceanic internal solitary waves. J Phys Oceanogr 44(4):1116–1132
Halpern D (1971) Observations of short-period internal waves in Massachusetts Bay. J Mar Res 29:116–132
Helfrich KR (1992) Internal solitary wave breaking and run-up on a uniform slope. J Fluid Mech 243:133–154
Helfrich KR, Melville WK (1986) On long nonlinear internal waves over slope-shelf topography. J Fluid Mech 167:285–308
Holloway PE, Pelinovsky EN, Talipova T, Barnes B (1997) A nonlinear model of internal tide transformation on the Australian North West Shelf. J Phys Oceanogr 27:871–896
Holloway PE, Pelinovsky EN, Talipova T, Barnes B (1999) A generalized Korteweg-de Vries model of internal tide transformation in the coastal zone. J Geophys Res 104:18333–18350
Ivey GN, Nokes RI (1989) Vertical mixing due to the breaking of critical internal waves on sloping boundaries. J Fluid Mech 204:479–500
Keller JB, Mov VC (1969) Internal wave propagation in an inhomogeneous fluid of non-uniform depth. J Fluid Mech 38(2):365–374
Krauss W (1966) Interne Wellen. Gebruder Borntraeger, Berlin
Kurkina O, Talipova T, Pelinovsky E, Soomere T (2011) Mapping the internal wave field in the Baltic Sea in the context of sediment transport in shallow water. J Coast Res 64:2042–2047 (Special issue)
Larsen LH (1969) Internal waves incident upon a knife edge barrier. Deep-Sea Res 16:411–419
Lien RC, Henyey F, Ma B (2014) Large-amplitude internal solitary waves observed in the Northern South China Sea: properties and energetics. J Phys Oceanogr 44(4):1095–1115
Massel SR (1989) Hydrodynamics of coastal zones. Elsevier, Amsterdam
Massel SR (2007) Ocean wave breaking and marine aerosol fluxes. Springer, New York
Melnikov WA (1982) Influence of the bottoms form on the internal waves. Fizika Atm i Okeana 18:775–778 (in Russian)
Miropolsky YZ (1974) Propagation of internal waves in ocean with horizontally inhomogeneous density field. Fizika Atm i Okeana 10(5) (in Russian)
Müller P, Liu X (2000a) Scattering of internal waves at finite topography in two dimensions. Part I: theory and case studies. J Phys Oceanogr 30:532–549
Müller P, Liu X (2000b) Scattering of internal waves at finite topography in two dimensions. Part II: spectral calculations and boundary mixing. J Phys Oceanogr 30:550–563
Osborne AR, Burch TL (1980) Internal solutions in the Andaman Sea. Science 208(4443):451–460
Ostrovsky LA (1978) Nonlinear internal waves in a rotating ocean. Oceanologiya 18(2):119–125 (in Russian)

Pedlosky J (2003) Waves in the ocean and atmosphere. Introduction to wave dynamics. Springer, Berlin
Pelinovsky EN, Shavratsky SCh (1976) Propagation on nonlinear internal waves in an inhomogeneous ocean. Fizika Atm i Okeana 12:76–82 (in Russian)
Pelinovsky EN, Talipowa TG, Stepanyants JA (1994) Modelling of nonlinear wave propagation in the horizontally inhomogeneous ocean. Fizika Atm i Okeana 30(1):79–83 (in Russian)
Phillips OM (1966) The dynamics of the upper ocean. Cambridge University Press, Cambridge
Robinson RM (1969) The effects of a vertical barrier on internal waves. Deep-Sea Res 16:421–429
Sandström H (1969) Effect of topography on propagation of waves in stratified fluids. Deep-Sea Res 16:405–410
Smyth NF, Holloway PE (1988) Hydraulic jamp and undular bore formation on a shelf break. J Phys Oceanogr 18:947–962
Sutherland B (2010) Internal gravity waves. Cambridge University Press, Cambridge
Talipova T, Pelinovsky EN, Kouts T (1998) Kinematic characteristics of internal wave field in the Gotland deep of the Baltic Sea. Fizika Atm i Okeana 38:37–46 (in Russian)
Thorpe SA (1987) On the reflection of a train of finite amplitude internal waves from a uniform slope. J Fluid Mech 178:279–302
Thorpe SA (1997) On the interactions of internal waves reflecting from slopes. J Phys Oceanogr 27:2072–2078
Thorpe SA (2005) The turbulent ocean. Cambridge University Press, Cambridge
Vlasenko V, Hutter I (2002) Numerical experiments on the breaking of solitary internal waves over a slope-shelf topography. J Phys Oceanogr 32:1779–1793
Whitham GB (1974) Linear and nonlinear waves. Wiley, New York
Wunsch C (1968) On the propagation of internal waves up a slope. Deep-Sea Res 15:251–258
Wunsch C (1969) Progressive internal waves on slopes. J Fluid Mech 35(1):131–144

References

Chapter 5
Generation of Internal Waves

5.1 Introduction

The mechanisms of generation of internal waves in the stratified fluid are not exactly known and they are still a subject of observation and numerical modelling. Apart from the tides flowing over topography, the other driving force generating internal waves results are fluctuations of atmospheric pressure and sudden impact of stormy winds. Moreover, the bodies oscillating with frequency ω in the stratified fluid, with buoyancy frequency N_0, create a cross-shaped pattern of internal waves when $\omega < N_0$. However, this generation mechanism is rather far from the geophysical flows discussed in this book, and it will not be considered here. A comprehensive information on the internal waves generated by solid oscillating bodies is given in a book by Sutherland (2010).

In Sect. 5.2, the interaction of the stratified flow with local disturbances as well as with the continental shelf and bottom irregularities is considered. The complicated processes of the energy transfer from the local winds to the inertial motions is described in Sect. 5.3. The influence of basic atmospheric and hydrological parameters such as wind stress, water stratification and turbulent dissipation on the observed near-inertial waves is discussed.

5.2 Generation of Internal Waves by Localised Disturbances

Theoretical investigations and experiments show that interaction of the tidal currents (barotropic tidal flows) with underwater topographic features, like continental slopes, sills and underwater banks, generates the most energetic internal waves.

The simplest case of the interaction of flow with topography is a disturbance generated by the flow over localised small-amplitude hills. The standard Fourier transform method can be used to predict the structure of waves generated by a hill of an arbitrary shape. Let the hill height be prescribed by $h(x)$ when $h(x)k_z \ll 1$.

S.R. Massel, *Internal Gravity Waves in the Shallow Seas*,
GeoPlanet: Earth and Planetary Sciences, DOI 10.1007/978-3-319-18908-6_5

Therefore Fourier transform of $h(x)$ is (Sutherland 2010):

$$\hat{h}(x) = \frac{1}{2\pi} \int_{-\infty}^{\infty} h(x) e^{-ik_x x} dx \tag{5.1}$$

The vertical displacement field of waves generated by the hill becomes:

$$\zeta(x, z) = \int_{-\infty}^{\infty} \hat{h} e^{i(k_x x + k_z z)} dk_x \tag{5.2}$$

when wavenumber k_z is a function of k_x (see Chap. 3).

The explicit analytic solutions of Eq. (5.1) generally cannot be found, however, the results can be obtained numerically using the FFT method. Some exception for the analytic solution is an isolated hill having the 'Witch of Agnesi' shape given by:

$$h(x) = A\left[1 + \left(\frac{x}{L}\right)^2\right]^{-1} \tag{5.3}$$

where L is the extent of the hill. Sutherland (2010) argues that when Froude number $Fr = U/N_0 L \ll 1$ (U is the indent flow velocity), vertically propagating waves with wavelengths of the scale of the hill are generated. If $Fr = U/N_0 L \gg 1$, the waves which are very long compared to the extent of the hill are observed.

Miles and Huppert (1968) determined the wave pattern for stratified flow over an obstacle-semi-circular hill of radius R_0. Vertically propagating internal waves are more clearly seen for smaller non-dimensional Froude number $Fr = U/N_0 R_0$ and vertical displacements of isopycnals become more pronounced then. For the high speed of flow ($Fr \approx 2.0$), the obstacle excites evanescent disturbances over the hill with long, small-amplitude waves behind it.

Internal waves generated by stratified flow over large-amplitude localised hills were considered by Vlasenko et al. (2005).

Sutherland and Linden (1998) performed the laboratory and numerical experiments on the excitation of internal waves in the wake of a tall and thin vertical barrier. Large-amplitude internal waves have been generated when the mixing region is weakly stratified and the deep fluid is sufficiently strongly stratified. Internal waves are generated by vortices shed in the wake of the barrier. They should be distinguished from the excitation of lee waves which develop on a scale U/N_0.

Let us now consider the geophysical case of interaction of the flow with topography. The good example is a strong internal solitary wave packet generated by the interaction of the barotropic tide with shallow bottom topographic features located in the Andaman Sea (Dreadnought Bank). The ERS-2 SAR image of this area on 11 February 1997 showed sea surface manifestations of internal waves with some almost semicircular small-scale wave pattern of the secondary internal waves (Vlasenko and Alpers 2005).

To describe such a complicated pattern of the wave-topography interactions, fully numerical 3D models of baroclinic tides are usually applied (see for example Holloway and Merrifield 1999; Niwa and Hibiya 2001; Mahadevan et al. 1996; Marshall et al. 1997). An approximate solution to the complex 3D nonlinear and nonhydrostatic systems can be obtained using the advanced numerical 2D models. Such approach was employed by Vlasenko and Alpers (2005) with the aim to explain the observed process of generation of secondary waves by scattering of internal solitary waves at the Dreadnought Bank in the Andaman Sea. The good agreement between the model and remote sensing data supports the existence of this generation mechanism, taking into account that only the leading soliton of the incident wave packet generates the secondary wave packet.

A barotropic tide which forces the stratified water to oscillate over continental slope topography is often a source of long internal waves. Observations have been reported from various shelf and slope regions, for example the U.K. shelf, off the northwest African shelf, and the Australian North West Shelf (NWS) (Holloway 1983, 1996).

The internal tides transport, advect and disperse sediment particles shaping the sea bed through resuspension due to strong near-bed flows. The most favourable generation regions for internal tides are where the slope of the bottom β equals approximately the critical angle and the internal wave characteristics c (see Eq. 4.13). It is believed that internal tides with angles of 2°–4° are the reason for the universal 3° slopes of the continental shelfs. Models of internal tides propagation (Wunsch 1969) as well as observations (Holloway 1984) indicate that the currents near the sea bottom are strongly intensified close to the critical slope. Holloway (1983) calculated the wave characteristics c over the NWS using the temperature survey data for the M_2 tidal frequency. Everywhere the characteristics are steeper than the bottom slope, and there is no well-defined region of generation where the characteristics and bathymetry are parallel. The generation takes place over a very large section of the bottom and not from one point. Horn and Meincke (1976) found a similar intensification of the M_2 baroclinic tidal energy on the slope region off northwest Africa.

Further insight into the mechanism of generation of the internal tides is obtained from numerical models. Holloway (1996) applied a nonlinear, finite-difference model of the generation, propagation and dissipation of internal waves to the shelf region of the NWS when only the vertical and the cross-shelf variabilities are taken into account. The model was forced through the specification of the offshore tidal elevation with the M_2 constituent. It was shown that amplitude and phase of the vertical displacement of internal waves are dependent on the slope of the bathymetry with largest values occurring for steepest topography. In particular, the largest values of energy flux were confined to the regions close to the sea bottom and at locations with depths identified at regions of near-critical bottom slope and of internal wave generation. The energy flux also depends on the seasonal vertical structure of water column. The horizontal velocity profiles are changing significantly from the summer to the winter stratification, with strong bottom intensification of currents in winter.

Van Gastel et al. (2009) applied hydrostatic, primitive equation model solving the Reynolds averaged form of the Navier-Stokes equations. Tidal forcing included the

M2, S2, N2, K2, O1, K1, P1 and Q1 tidal constituents. Model revealed the 3D nature of internal waves generation at water depths of between 400 and 600 m, strongly depending on the seasonal variation.

5.3 Generation of Internal Waves by Wind Action and Atmospheric Pressure Fluctuations

The observations of upper ocean velocities show that fluctuations in the local wind stress induce oscillations at nearly the local frequency. Inertial motions may exchange energy with other oceanic motions. In particular, the mixed-layer inertial currents generated by wind propagate downward and excite inertial currents beneath the mixed layer. To compute the average flux of energy from the wind to the mixed layer of inertial currents at fixed depth h, the simple damped slab model of the mixed layer was applied by D'Asaro (1985). So, we consider a mixed layer of depth h, density ρ, inertial frequency f and wind stress components τ_x and τ_y. Therefore the velocity components become:

$$\left.\begin{aligned} \frac{du}{dt} - fv &= \frac{\tau_x}{h\rho} - ru \\ \frac{dv}{dt} + fu &= \frac{\tau_y}{h\rho} - rv \end{aligned}\right\} \tag{5.4}$$

where r is a damping constant that parameterizes the transfer of energy from the mixed layer to the deeper sea. The best estimate for r is such that $1/r \approx 2$–10 days.

If we introduce the following complex quantities as:

$$\left.\begin{aligned} Z &= u + iv \\ T &= \frac{\tau_x + i\tau_y}{\rho} \\ R &= r + if \end{aligned}\right\} \tag{5.5}$$

we obtain the model equations in the form:

$$\frac{dZ}{dt} + RZ = \frac{T}{h} \tag{5.6}$$

For the zero wind stress ($T = 0$), the solution of Eq. (5.6) is:

$$Z = Z_0 \exp(-Rt)\exp(-ift) \tag{5.7}$$

Velocity vector is changing with the frequency f and decays at the rate R, which corresponds to the damped inertial oscillations.

For steady wind, the solution of Eq. (5.6) represents the Ekman transport modified by frictional damping, i.e.:

$$Z_E = \frac{T}{Rh} \tag{5.8}$$

When the wind stresses are arbitrary, the solution is composed by the time varying inertial component and Ekman component. For the inertial oscillation component we have:

$$Z_I = Z - Z_E \tag{5.9}$$

Thus, from Eq. (5.6) we obtain:

$$\frac{dZ_I}{dt} + RZ_I = -\frac{dZ_E}{dt} \tag{5.10}$$

or

$$\frac{dZ_I}{dt} + RZ_I = -\frac{1}{Rh}\frac{dT}{dt} - \frac{T}{R}\frac{d(1/h)}{dt} \tag{5.11}$$

As the mixed-layer depth changes much slower than the wind stress changes, Eq. (5.11) can be simplified as follows:

$$\frac{dZ_I}{dt} + RZ_I = -\frac{1}{Rh}\frac{dT}{dt} \tag{5.12}$$

D'Asaro (1985) discussed the mixed layer inertial motions induced by winds measured at a number of middle latitude buoys maintained by NOAA Data Buoy Center. The values of τ are available at 1 or 3 h intervals and the value of parameter $1/r \approx 4$ days results in the most realistic simulated inertial currents.

When the wind stress at times t_1 and t_2, and inertial current Z_{I1} at t_1 are known, Eq. (5.10) can be solved as:

$$Z_{I2} = Z_{I1}\exp(-R\Delta t) - \frac{T_1}{hR^2}(1 - \exp(-R\Delta t)) \tag{5.13}$$

in which $T_1 = \Delta T/\Delta t$ represents the change of T over the interval Δt.

D'Asaro (1985) evaluated the average flux of energy transferred to the inertial motions during the interval Δt in the form:

$$F_E = Re\left\{\frac{T_I^*}{\Delta t|R|^2 h}\left[\left(Z_{I1} + \frac{T_I}{R^2 h}\right)(\exp(-R\Delta t) - 1) + \frac{T_I}{\omega h}\Delta t\right]\right\} \tag{5.14}$$

in which T_I^* is the complex conjugate of T_I and value of F_E is associated with time $t = t_2$.

Process of the inertial motion generation exhibits strong intermittency. Therefore, in the linear models, the wind stress is sometimes represented as a series of delta functions, each being an "inertial event". It should be noted that after a sudden onset of wind, the ocean response is initially dominated by inertial motion, and later, the ocean tends towards a state of a geostrophic equilibrium due to inertial waves radiating away from the source.

Inertial waves can only be generated if there are spatial inhomogenities, such as lateral boundaries. The waves are generated at the boundary and propagate away from them. Millot and Crepon (1981) observed inertial oscillations at the northwestern Mediterranean coast, and Anderson et al. (1983) recorded substantial inertial oscillations with the amplitudes of 20 cm s^{-1} off the Oregon coast. The Millot and Crepon's two-layer model indicates that the generated interface oscillations have frequencies higher than frequency f and propagate offshore at a speed close to the speed of long baroclinic waves:

$$C_0^2 = \frac{g(\rho_2 - \rho_1)h_1 h_2}{\rho_2(h_1 + h_2)} \tag{5.15}$$

The model suggests that the observed inertial oscillations are caused by two mechanisms, namely the local response of the ocean to meteorological forcing and the propagation of long internal waves generated in the transient phase of the geostrophic adjustment. Klinck et al. (1981) developed a linear model of a rotating, linearly stratified fluid to investigate the response of a flat-bottom coastal area to a moving surface wind stress. They showed that if the speed of the atmospheric forcing U is higher than the internal wave speed C_0, two sets of inertia waves are generated. One set is forced directly by the moving surface stresses and moves at the speed of the storm. The other set of waves results from reflection of the directly forced waves from the coastal wall, and radiates from the wall with slowly decreasing amplitude. The model was used to analyse the effects of cold fronts moving on a continental shelf, such as the South Atlantic Bight.

Kundu et al. (1983) developed an analytical linear solution for stratified, semi-infinite ocean with a flat bottom and straight coast when wind stress τ is suddenly imposed along the coast. As a layer of low stratification frequently exists near the surface, due to the previous storms, the stratification is taken to be zero at the surface of mixed layer of constant thickness h_0. The solution of the governing equations was obtained by expanding the variable in terms of the vertical normal modes, i.e.:

$$u, v, p = \sum_{n=0}^{\infty} (u_n, v_n, p_n)\phi_n \tag{5.16}$$

$$\rho = \sum_{n=0}^{\infty} \rho_n \frac{\partial \phi_n}{\partial z} \tag{5.17}$$

where:

$$\frac{\partial \phi_n}{\partial z} + \frac{N_0^2}{(C_0^{(n)})^2} \int_{-h}^{0} \phi_n dz = 0 \tag{5.18}$$

where N_0 is the uniform stratification of the fluid.

The eigenvalue $C_0^{(n)}$ is the phase speed of the long internal wave of mode n. Due to assumption of the rigid lid we have $C_0^{(n)} = N_0 h/(n\pi)$ (see Eq. 3.22 when the Earth rotation is neglected). For $C_0^{(n)} = \infty$, the barotropic mode is reached.

We assume that the constant wind stress $\tau^{(y)}$ is suddenly applied along shore, i.e. $\tau^{(y)} = \tau_0^{(y)} H(t)$, where $H(t)$ is the Heaviside unit function and $\tau_0^{(y)}$ is independent of x and t. The method of Laplace transform is used to solve the basic equations. The solution for inviscid motion for the first or second modes of the pressure and velocity components can be summarized as follows (Kundu et al. 1983):

$$p_n = H(\hat{t}) f \tau_n C_0^{(n)} \{t\} * \{J_0(\sqrt{\tilde{x}})\} \tag{5.19}$$

$$u_n = \frac{\tau_n}{f}[1 - \cos(ft)] - H(\hat{t})\tau_n f^{-1}[1 - \cos(ft + x_n)] - H(\hat{t})\tau_n x_n \{1 - \cos(ft)\} * \left\{\frac{J_1(\sqrt{\tilde{x}})}{\sqrt{\tilde{x}}}\right\} \tag{5.20}$$

$$v_n = \frac{\tau_n}{f}\sin(ft) - H(\hat{t})\tau_n f^{-1}[\sin(ft + x_n) - ft - x_n] - H(\hat{t})\tau_n x_n \{\sin(ft) - ft\} * \left\{\frac{J_1(\sqrt{\tilde{x}})}{\sqrt{\tilde{x}}}\right\} \tag{5.21}$$

$$w_n = H(\hat{t}) \frac{f\tau_n}{C_0^{(n)}} \{1\} * \{J_0(\sqrt{\tilde{x}})\} \tag{5.22}$$

where:

$$\hat{t} = t + \frac{x}{C_0^{(n)}}, \quad \tilde{x} = f^2 t^2 - x_n^2, \quad x_n = \frac{xf}{C_0^{(n)}} \tag{5.23}$$

where $J_0(x)$ and $J_1(x)$ are the Bessel functions of the zero and first order. The inner product of two functions, delineated by braces, represent the integral from $-x/C_0^{(n)}$ to t, i.e.:

$$\{f(t)\} * \{g(t)\} = \int_{-x/C_0^{(n)}}^{t} f(t - \alpha) g(\alpha) d\alpha \tag{5.24}$$

Formulas (5.19)–(5.22) are complicated. To get some insight into the problem of inertial oscillations, Kundu et al. (1983) developed the asymptotic forms when $ft \to \infty$ and x_n is fixed:

$$p_n \approx \frac{\tau_n C_0^{(n)}}{f}\left[ft \exp(x_n) - \left(\frac{2}{\pi ft}\right)^{1/2} \sin\left(ft + \frac{\pi}{4}\right)\right] \tag{5.25}$$

$$u_n \approx \frac{\tau_n}{f}\left[1 - \exp(x_n) + x_n\left(\frac{2}{\pi ft}\right)^{1/2} \cos\left(ft + \frac{\pi}{4}\right)\right] \tag{5.26}$$

$$v_n \approx \frac{\tau_n}{f}\left[ft \exp(x_n) - x_n\left(\frac{2}{\pi ft}\right)^{1/2} \sin\left(ft + \frac{\pi}{4}\right)\right] \tag{5.27}$$

$$w_n \approx \frac{\tau_n}{C_0^{(n)}}\left[\exp(x_n) - \left(\frac{2}{\pi ft}\right)^{1/2} \cos\left(ft + \frac{\pi}{4}\right)\right] \tag{5.28}$$

Numerical calculations showed that after the barotropic front passed, inertial oscillations in the mixed layer are replaced by slab-like motions with frequency $\omega \approx f$ affecting the whole water column. Subsequent arrival of higher-mode baroclinic wave front generated at the coast results in a small shift towards higher frequencies ($\omega > f$).

The asymptotic forms (5.25)–(5.28) showed that the inertial oscillations decay like $t^{-1}/2$, leaving behind an accelerating coastally trapped baroclinic jet and a steady upwelling circulation. Within the Rossby radius $Re = C_0^{(n)}/f \approx 8$ km, used in the numerical calculations, the inertial oscillations decay within a matter of days. The energy flux from the surface layer into the deeper layer is accompanied by the upward phase propagation and the rapid decay of the oscillations near the coast at all depths.

To exhibit the periodic and aperiodic components of the solutions, Kundu et al. (1983) numerically evaluated the formulae (5.25)–(5.28) for total water depth of 250 m and depth of surface mixed layer $h = 25$ m. Uniform stratification $N = 10^{-2}\,\mathrm{s}^{-1}$ and wind stress $\tau^{(y)} = -10^{-1}\,\mathrm{Nm}^{-2}$ were applied for calculations. For these initial values, it was found that the oscillations decay near the coast at all depths. The inertial oscillations have started everywhere in the water column, with velocity amplitudes of 3.1 cm s^{-1} in the mixed layer and 0.4 cm s^{-1} below this layer. The amplitude of the inertial oscillations increases at the layer offshore distance of 32 km.

Kundu (1986) subsequently used a linear two-dimensional model for continuously stratified fluid to demonstrate the combined effect of the coastline forcing and the superposition of induced inertial oscillations.

Tintoré et al. (1995) showed that in the presence of a coastal boundary a transient wind forcing induces the offshore propagation of a wave front. After the passing of the front, a near inertial motion appears as a result of the downward propagation of internal wave energy. After several inertial periods, the kinetic energy in the upper layer can be removed. From the three-dimensional general circulation model with turbulence-closer mixed layer it was found that for the initial density profile, typical for the summer conditions off the northeast Spanish coast, an averaged Väisälä-Brunt frequency $N^2 \approx 3.5 \times 10^{-4}\,s^{-2}$. The resulting horizontal and vertical phase velocities are $40\,s^{-1}$ and $0.1\,s^{-1}$, respectively. Therefore, from formulae given in Chap. 3, the wave frequency $\omega \approx 1.06 \times 10^{-4}\,s^{-1}$ ($T \approx 16.5\,h$), while the local inertial frequency $f = 9.5 \times 10^{-5}\,s^{-1}$ (inertial period $T = 18.29\,h$). For the horizontal and vertical wavenumber we obtain, respectively, $k_x = 2.65 \times 10^{-6}\,cm^{-1}$ (wavelength $L_x = 24\,km$) and $k_z = 1.06 \times 10^{-3}\,cm^{-1}$ (wavelength $L_x = 59\,m$).

Fennel (1989) discussed the inertial response of a stratified basin of constant water depth with straight walls to longshore wind. He pointed out that in a closed basin Poincaré waves may be reflected from adjacent coasts, which leads to a complex superposition of modes. After reflection, the dispersive properties of waves are conserved and resulting stochastic patterns are characterized by locally varying frequencies and wavenumbers. Close behind the wave fronts which move with the maximum group velocity, the frequencies and wavenumbers are relatively high. Far behind the fronts, the frequency approaches the local inertial frequency, f, and the wavenumber becomes smaller. The theoretical results are consistent with observations in the Baltic Sea and in Lake Michigan.

In general, behaviour of the internal waves on the continental shelf depends on the basic atmospheric and hydrological parameters such as surface wind stress, τ, stratification, N, and turbulent dissipation, ε. MacKinnon and Gregg (2005) observed near-inertial internal waves on the New England Shelf, when stratification was rapidly changing. As the stratification evolved, the vertical structure of these near-inertial waves changed from mode 1 to mode 2. Growth of mode 2 was due to action of changing wind stress forcing and a nonlinear coupling between the first and second vertical modes. Two dimensional slab mixed layer model reproduced the magnitude and phase of the observed near-inertial waves. In particular, it successfully captures the increasing of the strength of mode-2 waves due to the transfer of energy between mode-1 and mode-2 waves through bottom stress.

References

Anderson I, Huyer A, Smith RL (1983) Near-inertial motions off the Oregon coast. J Geophys Res 88:5960–5972

D'Asaro EA (1985) The energy flux from the wind to near-inertial motions in the surface mixed layer. J Phys Oceanogr 15(9):1043–1059

Fennel W (1989) Inertial waves and inertial oscillations in channels. Cont Shelf Res 9(5):403–426

Holloway PE (1983) Internal tides on the Australian North-West Shelf: a preliminary investigation. J Phys Oceanogr 14:1778–1790

Holloway PE (1984) On the semi-diurnal internal tide at a shelf-break region on the Australian North West Shelf. J Phys Oceanogr 14:1778–1790

Holloway P (1996) A numerical model of internal tides with application to the Australian North West Shelf. J Phys Oceanogr 26(1):21–37
Holloway PE, Merrifield MA (1999) Internal tide generation by seamounts, ridges and islands. J Geophys Res 104:25937–25951
Horn W, Meincke J (1976) Note on the tidal current field in the continental slope area off northwest Africa. Mem Soc R Sci Liege 10:31–42
Klinck JM, Pietrafesa LJ, Janowitz GS (1981) Continental shelf circulation induced by a moving, localized wind stress. J Phys Oceanogr 11:836–848
Kundu PK (1986) A two-dimensioanl model of inertial oscillations generated by a propagating wind field. J Phys Oceanogr 16:1399–1411
Kundu PK, Chao S-Y, McCreary JP (1983) Transient coastal currents and inertio-gravity waves. Deep-Sea Res Part A 30(10):1059–1082
MacKinnon JA, Gregg MC (2005) Near-inertial waves on the New England shelf: the role of evolving stratification, turbulent dissipation, and bottom drag. J Phys Oceanogr 35(12):2408–2424
Mahadevan A, Oliger J, Street R (1996) A non-hydrostatic mesoscale ocean model. Part 2: numerical implementation. J Phys Oceanogr 26:1881–1900
Marshall J, Adcroft A, Hill C, Perelman L, Heisey C (1997) A finite-volume, incompressible Navier-Stokes model for studies of the ocean on parallel computers. J Geophys Res 102:5753–5766
Miles JW, Huppert HE (1968) Lee waves in a stratified flow. Part 2. Semi-circular obstacle. J Fluid Mech 33(4):803–814
Millot C, Crepon M (1981) Inertial oscillations on the continental shelf of the Gulf of Lions—observations and theory. J Phys Oceanogr 11(5):639–657
Niwa Y, Hibiya T (2001) Numerical study of the spatial distribution of the M_2 internal tide in the pacific ocean. J Geophys Res 106:22441–22449
Sutherland BR (2010) Internal gravity waves. Cambridge University Press, Cambridge. 377 p
Sutherland BR, Linden PF (1998) Internal wave excitation from stratified flow over a thin barrier. J Fluid Mech 377:223–252
Tintoré J, Wang D-P, Garcia E, Vińdez A (1995) Near-inertial motions in the coastal ocean. J Mar Syst 6:301–312
Van Gastel P, Ivey GN, Meuleners MJ, Antenucci JP, Fringer O (2009) The variability of the large-amplitude internal wave field on the Australian North West Shelf. Cont Shelf Res 29:1373–1383
Vlasenko V, Alpers W (2005) Generation of secondary internal waves by the interaction of an internal solitary wave with an underwater bank. J Geophys Res 110:C02019
Vlasenko V, Stashchuk N, Hutter K (2005) Baroclinic tides. Theoretical modelling and observational evidence. Cambridge University Press, Cambridge. 351 p
Wunsch C (1969) Progressive internal waves on slopes. J Fluid Mech 35(1):131–144

Chapter 6
Spectral Properties of Internal Waves

6.1 Introduction

Till now we focused primarily upon plane waves or quasi monochromatic wave packets. However, the internal waves in the ocean are generated with a variety of spatial end temporal scales, obeying frequencies as low as the inertial frequency f, up to frequency as high as the Väisälä-Brunt frequency, N. The observed cascade energy along the frequencies results from the large-scale internal wave generation mechanisms such as winds and tides, through an internal wave continuum to turbulence.

From discussion in the preceding chapters it follows that the internal waves have a modal vertical structure which influences the spectral distribution of internal waves. In Sect. 6.2, the general methodology of the spectral representation, in frequency and wavenumber space, for internal waves with a given mode structure is described. For experimental data, stationary in time or uniform in space, the Fourier Integral or Fourier-Stjeltjes Integral methods are available. However, the experimental data are usually non-stationary or non-uniform, and the information on the evolution of the spectra in time or space is required. This can be given by application of the Wavelet Transform or Hilbert Transform techniques.

In the open ocean, far from the continental shelves and coastlines, the ensemble of internal waves is found to be statistically homogeneous in time and space. This fact provides a synthesized spectrum formula known as the Garrett-Munk spectrum (Garrett and Munk 1975), discussed in Sect. 6.3 in detail.

The application of the spectral methodology for the shallow water is given in Chap. 7 when the experimental data collected in the Baltic Sea is used to illustrate spectral features of the internal waves.

The statistical methods are a natural supplement of the spectral method for random time series. However, the papers on the statistics of the internal wave are not numerous in a professional literature. To estimate the frequency of appearance of large-amplitude internal waves, Ivanov et al. (1993) showed that number n of internal waves with height bigger than H observed in the period T can be estimated by the Poisson type distribution:

S.R. Massel, *Internal Gravity Waves in the Shallow Seas*,
GeoPlanet: Earth and Planetary Sciences, DOI 10.1007/978-3-319-18908-6_6

$$n = NT \exp\left(-\frac{H}{\bar{H}}\right) \tag{6.1}$$

where $\bar{H}$ is a mean wave height and N is the Väisälä-Brunt frequency. Relationship agrees well with observations for $\bar{H} \approx 3.4$ m and $N =$ cycles/hour. In particular, the three-day experiments close to the Egiptian coast showed that 99 waves had a height of 0.1–0.5 m, 16 waves were of height 1.0–1.5 m and only 2 waves were as height as 2.0–2.5 m (Ivanov et al. 1993). Using the relationship (6.1), we can predict the internal wave heights H_T which can appear in this region in a given time T. Thus, for example, we obtain $H_T = 3.6$ m/week, $H_T = 4.5$ m/month and $H_T = 5.1$ m/three months.

Results of similar observations were made close to the Israel coast and on the tidal ocean region of the Australian North West Shelf (Pelinovsky et al. 1995). The semidiurnal internal tides play an important role in generation of short-scale waves exceeding particular values. Measurement of current fluctuations due to short period internal waves has been analysed with a view of calculating the exceedance probability of intense internal waves using the Poisson statistics.

6.2 Internal Waves Spectra with a Given Modal Structure

Integral representations of some characteristic parameters of the internal waves give a possibility to develop corresponding spectral functions in quite general forms. First, we use the Stjeltjes-Fourier integral technique to determine the frequency spectra of wave velocities and isopycnal displacements. For simplicity we restrict our discussion to the internal waves propagating in the vertical plane (x, z) with an exponential density profile and constant Väisälä-Brunt frequency N.

Let us first consider an analytical solution of the vertical velocity given in Chap. 3. The simplest and the most natural representation of the confused internal waves at a given level z would be the linear superposition of many harmonics ω and modes n. Therefore the deterministic representation of vertical velocity under the exponential density distribution becomes (see Eq. 3.30):

$$w(x, z, t) = \Re \sum_l \sum_n B_l^{(n)} \sin\left(\frac{n\pi z}{h}\right) \exp\left[i\left(k_{l,x}^{(n)} x - \omega_l t\right)\right] \tag{6.2}$$

and the Fourier-Stjeltjes integral representation of function $w(x, z, t)$ for a given point $x = 0$ is:

$$w(z, t) = \Re \sum_n \int_\omega \Pi_w^{(n)}(\omega, z) \exp(-i\omega t) dB^{(n)}(\omega) \tag{6.3}$$

where:

$$\Pi_w^{(n)}(\omega, z) = \sin\left(\frac{n\pi z}{h}\right) \tag{6.4}$$

By definition, the autocorrelation function becomes:

$$K_w(\tau; z, z) = \overline{w(z, t)w(z, t + \tau)} \tag{6.5}$$

in which an overline denotes the averaging in the stochastic sense.

Thus, from Eq. (6.3) we have:

$$K_w(\tau; z, z) = \Re \sum_n \sum_m \int_\omega \int_{\omega_1} \overline{\Pi_w^{(n)}(\omega, z)\left(\Pi_w^{(m)}(\omega_1, z)\right)^*} \times \exp(i\omega\tau)\, \overline{dB^{(n)}(\omega) dB^{(m)}(\omega_1)} \tag{6.6}$$

where (∗) denotes the conjugate of the complex variable.

The stationarity condition requires that:

$$\overline{dB^{(n)}(\omega) dB^{(m)}(\omega_1)} = S_w^{(n)}(\omega)\delta(\omega - \omega_1)\delta_{nm}\, d\omega\, d\omega_1 \tag{6.7}$$

in which $\delta(\omega - \omega_1)$ is the Dirac's delta:

$$\int_{-\infty}^{\infty} f(x_1)\delta(x - x_1)dx_1 = f(x) \tag{6.8}$$

and δ_{nm} is the Kronecker's delta when:

$$\delta_{nm} = \begin{cases} 1 \text{ for } n = m \\ 0 \text{ for } n \neq m \end{cases} \tag{6.9}$$

The function $S_w^{(n)}(\omega)$ is the frequency spectrum of the vertical velocity for given mode n.

Substituting (6.7) into (6.6) we obtain:

$$K_w(\tau; z, z) = \sum_n \int_\omega \overline{\Pi_w^{(n)}(\omega, z)\left(\Pi_w^{(n)}(\omega, z)\right)^*} \exp(i\omega\tau) S_w^{(n)}(\omega) d\omega \tag{6.10}$$

When $\tau = 0$, the autocorrelation function denotes the variance of the random function $w(z, t)$. Thus:

$$\sigma_w^2(z) = \sum_n \int_\omega \overline{\Pi_w^{(n)}(\omega, z)\left(\Pi_w^{(n)}(\omega, z)\right)^*} S_w^{(n)}(\omega) d\omega \tag{6.11}$$

Therefore, the final frequency spectrum of the vertical velocity at level z becomes:

$$S_w(\omega, z) = \sum_n \overline{\Pi_w^{(n)}(\omega, z)\left(\Pi_w^{(n)}(\omega, z)\right)^* S_w^{(n)}(\omega)} \quad (6.12)$$

In particular, for exponential density distribution, function $\Pi_w^{(n)}(\omega, z)$ is given by Eq. (6.4), and the frequency spectrum for vertical velocity takes a form:

$$S_w(\omega, z) = \sum_n \sin^2\left(\frac{n\pi z}{h}\right) S_w^{(n)}(\omega) \quad (6.13)$$

In a similar way, we represent the isopycnal displacements ζ in the form (see Eq. 3.32):

$$\zeta(x, z, t) = \Re \sum_l \sum_n \frac{B_l^{(n)}}{\omega_l} \sin\left(\frac{n\pi z}{h}\right) \exp\left[i\left(k_{l,x}^{(n)} x - \omega_l t + \frac{\pi}{2}\right)\right] \quad (6.14)$$

Therefore, using the same procedure as above we obtain the final frequency spectrum of the isopycnal displacements at level z as follows:

$$S_\zeta(\omega, z) = \sum_n \overline{\Pi_\zeta^{(n)}(\omega, z)\left(\Pi_\zeta^{(n)}(\omega, z)\right)^* S_w^{(n)}(\omega)} \quad (6.15)$$

For for exponential density distribution we have:

$$\Pi_\zeta^{(n)}(\omega, z) = \frac{i}{\omega} \sin\left(\frac{n\pi z}{h}\right) \quad (6.16)$$

Thus the corresponding frequency spectrum is:

$$S_\zeta(\omega, z) = \sum_n \frac{\sin^2\left(\frac{n\pi z}{h}\right)}{\omega^2} S_w^{(n)}(\omega) \quad (6.17)$$

It should be noted that the frequency spectra $S_w^{(n)}(\omega, z)$ and $S_\zeta^{(n)}(\omega, z)$ satisfy the condition that $w^{(n)}(t) = \partial\zeta(t)/dt$ and therefore the corresponding frequency spectra for a given mode n are:

$$S_w^{(n)}(\omega) = \omega^2 S_\zeta^{(n)}(\omega) \quad (6.18)$$

Monin and Krasitskii (1985) developed the expressions for the wavenumber spectra of the internal waves using the Fourier integrals. In particular, for the isopycnal displacements we can write:

$$\zeta(\mathbf{x}, z, t) = \sum_n \int_{\mathbf{k}} \Pi_\zeta^{(n)}(z) \exp\left[i\left(\mathbf{kx} - \omega_{\mathbf{k}}^{(n)} t\right)\right] A_{\mathbf{k}}^{(n)} d\mathbf{k} \quad (6.19)$$

in which:

$$2\,\overline{A_{\mathbf{k}}^{(n)}\left(A_{\mathbf{k}_1}^{(m)}\right)^*} = F^{(n)}(\mathbf{k})\delta(\mathbf{k}-\mathbf{k}_1)\delta_{nm} \qquad (6.20)$$

Therefore the final wavenumber spectrum becomes:

$$F(\mathbf{k}; z, z) = \sum_n \overline{\Pi_\zeta^{(n)}(z)\left(\Pi_\zeta^{(n)}(z)\right)^*} F^{(n)}(\mathbf{k}) \qquad (6.21)$$

The formulae above will be used in Chap. 7 for description of the internal wave kinematics in the Baltic Sea.

In general, empirical spectral density of the isopycnal displacements (or isotherm displacements) can be obtained from the time series $\zeta(t)$ of some finite duration, $T = N\Delta t$, where Δt is the time sampling. If the time series is stationary, the discrete Fourier transform gives:

$$Y_k(f) = \Delta t \sum_{n=1}^{N} \zeta_n \exp\left[-\frac{2\pi i n}{N}\right] \qquad (6.22)$$

The Y_k values are given for frequencies $f_k = k/N\Delta t$, $k = 0, \ldots, N$. They are confined to the Nyquist internal $-f_N < f_k < f_N = 1/(2\Delta t)$. Since $f_{N-k} = f_k$, only the first $N/2$ Fourier Transform values are unique (Emery and Thomson 1997).

The one-side power spectral density $S(\omega)$ for the positive frequency interval only is:

$$S(\omega) = \frac{2}{N\Delta t}\left|Y(f)\right|^2 \qquad (6.23)$$

The Fourier Transform, shown above, yields information on how much of the particular frequency components exist, but not when (in time). Such information is sufficient in a case of stationary signals as the frequency signals does not change in time and all frequency components exist all the time. When the time localisation of the spectral components is required, the transform of time series which provides the time-frequency representation of the signal should be developed. A transform of such type is the Wavelet Transform which gives full time-frequency representation of the time series (Torrence and Compo 1998).

In general, the Wavelet Transform (WT) of the signal, $x(t)$, is defined as a following inner product:

$$WT(\tau, b) = < g_{\tau b} \mid x > = \int_{-\infty}^{\infty} x(t) g_{\tau b}^*(t; \tau, b)dt \qquad (6.24)$$

The family of continuously translated and dilated wavelets is generated from mother wavelet $g(t)$:

$$g_{\tau b}(t;\tau,b) = \frac{1}{\sqrt{b}}\, g\left(\frac{t-\tau}{b}\right) \tag{6.25}$$

where τ is the translation parameter, corresponding to the position of the wavelet as it is shifted through the signal, b is the scale dilation parameter determining the width of the wavelet and the (*) indicates the complex conjugate. The scale $b > 1$ dilates (or stretches out) the signals, whereas scale $b < 1$ compresses the signal. The wavelet coefficients, $WT(\tau, b)$, represent the correlation (in term of the time-scale functions) between the wavelet and a localised section of the signal. If the signal has a major component of the frequency corresponding to the given scale, then the wavelet at this scale is close to the signal at the particular location and the corresponding wavelet transform coefficient, determined at this point, has a relatively large value. Therefore, the Wavelet Transform is a sort of a *microscope* with magnification $1/b$ and location given by parameter τ, while the optics of the microscope is characterised by the function $g_{\tau b}(t;\tau,b)$. For the wavelet which has the mother wavelet status, the function $g(t)$ must be regular function such that it decays to zero in the limit $|t| \to \infty$ and it has zero mean (Massel 2001).

It should be noted that in wavelet analysis the signal is multiplied with a window (wavelet). However, the window width has been changed as the transform is computed for every single spectral component. Changing the window width influences the resolution of the transform (Massel 2001). At low frequency (high value of scale b), the frequency resolution is better but time resolution is poor (more ambiguity regarding the value of the exact time τ). On the other hand, at higher frequencies (low scale b), the height of the boxes increases and the width decreases. This corresponds to a very narrow window when the frequency resolution is getting poorer and the time resolution better.

The Wavelet Transform should reflect the type of features which are present in the time series. For time series with sharp steps, a boxcar-like wavelet should be chosen, while for smoothly varying time series a smooth function is more appropriate. However, if the wavelet power spectra are not of primary interest, the choice of wavelet function is not critical.

One of the most extensively used mother wavelets is the Morlet's wavelet:

$$g(t) = \exp\left(-\frac{1}{2}t^2\right)\exp(ict) \tag{6.26}$$

Equation (6.26) represents a plane wave of frequency c, modulated by a Gaussian envelope of the unit width. Using the representation (6.26), the Morlet wavelet takes the form (Massel 2001):

$$g_{\tau b}(t) = \frac{1}{\sqrt{b}}\exp\left[-\frac{1}{2}\left(\frac{t-\tau}{b}\right)^2\right]\exp\left[ic\,\frac{t-\tau}{b}\right] \tag{6.27}$$

The final formula for the Wavelet Transform now becomes:

$$WT(\tau, b) = \int_{-\infty}^{\infty} x(t) \frac{x(t)}{\sqrt{b}} \exp\left[-\frac{1}{2}\left(\frac{t-\tau}{b}\right)^2\right] \exp\left[i \frac{2\pi}{b}(t-\tau)\right] dt \qquad (6.28)$$

Generally, in oceanography we are dealing with the finite-length time series. Thus, some errors occur at the beginning and the end of the wavelet spectrum. These regions of the spectrum in which edge effects become important are known as the cone of influence.

In order to make some insight into the effects of the finite-length time series, let us consider the time series of length T. For arbitrary function $x(t)$, the Wavelet Transform can be determined only by the numerical integration. Due to final length of the time series, the first value of the translation parameter τ, corresponding to the position of the wavelet, should be selected such as:

$$\exp\left[-\frac{1}{2}\left(\frac{t-\tau}{b}\right)^2\right] = \exp\left[-\frac{1}{2}\left(\frac{\Delta t_1}{b}\right)^2\right] = \delta \ll 1.0 \qquad (6.29)$$

Thus:

$$\Delta t_1 = \sqrt{-2\ln\delta}\, b = \frac{2\pi\sqrt{-2\ln\delta}}{f_b} \qquad (6.30)$$

where $f_b = 2\pi/b$.

The same is applied to the last value of the translation value τ. In this way, the total cone width becomes:

$$\Delta t = \frac{4\pi\sqrt{-2\ln\delta}}{f_b} \qquad (6.31)$$

and

$$\Delta t f_b = 4\pi\sqrt{-2\ln\delta} \qquad (6.32)$$

The values of Δt and f_b are the width and height of the resolution boxes in the time-frequency plane. Let us assume that the accuracy $\delta = 10^{-3}$. Therefore $\Delta t f_b = 46.684$. It means that for scale dilation parameter $b = 15$ min (high frequency $f_b = 2\pi/15$), the first value of the translation parameter τ should be equal $\tau \approx 55.7$ min and the last one be $(T-\tau)$. It should be noted that the final width of the cone of influence equals $\Delta t = 111.51$ min. Thus, the Wavelet Transform can be determined only for the time extend $(T - \Delta t)$. On the other hand, for higher value of $b = 420$ min (low frequency $f_b = 2\pi/420$), the first value of τ should appear at $\tau = 1561$ min and the total width $\Delta t = 3122$ min. The time extend for which the Wavelet Transform can be determined $(T - \Delta t)$ will be much smaller.

For higher accuracy, from Eq. (6.32) we obtain: $\Delta t f_b = 53.907$ for $\delta = 10^{-4}$ and $\Delta t f_b = 60.269$ for $\delta = 10^{-5}$.

It can be shown that Wavelet Transform conserves the energy, i.e.:

$$\int_{-\infty}^{\infty} |x(t)|^2 \, dt = C^{-1} \int_0^{\infty} \int_0^{\infty} \mid WT(\tau, b) \mid^2 b^{-2} d\tau db \tag{6.33}$$

in which:

$$C^{-1} = \int_{-\infty}^{\infty} \left(\omega^{-1} |G(\omega)|^2 \right) d\omega \tag{6.34}$$

and the $G(\omega)$ is the Fourier Transform of function $g(t)$.

Using the Wavelet Transform and the coefficient C we can define various wavelet energy spectra and spectral densities. In particular, so called *time-scale energy density* becomes:

$$E_1(\tau, b) = \frac{\mid WT(\tau, b) \mid^2}{b} \tag{6.35}$$

By integrating Eq. (6.35) versus scale b we obtain the *local energy density*:

$$E_2(\tau) = C^{-1} \int_0^{\infty} E_1(\tau, b) \frac{db}{b} \tag{6.36}$$

On the other hand, the integration of Eq. (6.35) versus time τ gives the *global wavelet energy spectrum* $E_3(b)$ as:

$$E_3(b) = \int_0^{\infty} E_1(\tau, b) d\tau \tag{6.37}$$

Finally, the *total energy* of the time series $x(t)$ becomes:

$$E = C^{-1} \int_0^{\infty} E_3(b) \frac{db}{b} \tag{6.38}$$

The applications of the Wavelet Transform to the internal waves is illustrated in Chap. 7 for the wave records in the Southern Baltic.

As was shown above, the wavelet technique, based on Fourier spectral methodology with adjustable frequency-dependent window functions—the mother wavelets, provides temporal-frequency resolution for non-stationary signals. However, the Fourier methodology interprets wave non-linearity in terms of harmonic generation with the spectral energy leaking into the higher-frequency components. On the other hand, the Hilbert Transform interprets wave non-linearity as frequency modulation, and the spectral energy remains near the base frequencies.

To determine the function $\xi(t)$, let us assume for a moment that a real-valued function $\zeta(t)$ extends over the range $-\infty < t < \infty$. Then its Hilbert Transform becomes (Bendat and Piersol 1986):

$$\xi(t) = H[\zeta(t)] = P \int_{-\infty}^{\infty} \frac{\zeta(t)}{\pi(t - t_1)} dt_1 \qquad (6.39)$$

in which P denotes the Cauchy principle value. Equation (6.39) indicates that $\xi(t)$ is the convolution integral of $\zeta(t)$ and $1/\pi t$.

If the signal $\zeta(t)$ is represented in the form of a series, we can write:

$$\zeta(t) = \sum_{n=0}^{\infty} [a_n \cos(n\omega t) + b_n \sin(n\omega t)] \qquad (6.40)$$

and

$$\xi(t) = \sum_{n=0}^{\infty} [a_n \sin(n\omega t) - b_n \cos(n\omega t)] \qquad (6.41)$$

The digital computation of the Hilbert Transform was explained by Bendat and Piersol (1986) and Massel (2007), where it was also demonstrated that the Hilbert Transform of the velocity potential function appears to be the stream function of the water motion.

Combining the real-valued function $\zeta(t)$ with its Hilbert Transform $\xi(t)$ we obtain a complex analytical signal in the time domain:

$$\eta(t) = \zeta(t) + i\xi(t) = A(t) \exp(i\theta(t)) \qquad (6.42)$$

in which $A(t)$ is interpreted as the instantaneous amplitude:

$$A(t) = \sqrt{\zeta^2(t) + \xi^2(t)} \qquad (6.43)$$

and $\theta(t)$ is the instantaneous phase:

$$\theta(t) = \arctan\left(\frac{\xi(t)}{\zeta(t)}\right) \qquad (6.44)$$

Using the Hilbert Transform it is possible to derive the instantaneous frequency $\omega(t)$ in a time series. From Eq. (6.42) we have:

$$\zeta(t) = \Re(\eta(t)) = A(t) \cos(\theta(t)) \qquad (6.45)$$

Therefore instantaneous frequency ω is given by:

$$\omega = \frac{d\theta(t)}{dt} \tag{6.46}$$

However, a straightforward application of the Hilbert Transform sometimes leads to the appearance of positive and negative frequency values. To overcome this difficulty, Huang et al. (1998) introduced the empirical mode decomposition method. They found that at any given time, there is only one frequency value which can only represent one component. To satisfy this requirement, some limitations on the data are necessary. Probably the most obvious restriction is the assumption that the process under consideration is of narrow band. However, this is usually not true for the internal waves records.

The Hilbert Transform also provides the instantaneous velocities of the internal waves as follows:

– vertical velocity at level z

$$w(t; z) = \frac{\partial\zeta(t)}{\partial t} \tag{6.47}$$

– horizontal velocity at level z

$$u(t; z) = \frac{\partial\xi}{\partial t} \tag{6.48}$$

6.3 Experimental Spectra for Internal Waves in Deep Ocean

Experiments in deep ocean, away from continents and solid boundaries, showed that the ensembles of waves are statistically homogeneous in time and space. The observations do not show a random distribution of internal waves in the ocean, but rather their average spectra in the open sea. Garrett and Munk (1975, 1979) and Munk (1981) proposed the empirical reference wavenumber and frequency spectra, labelled GM75 and later GM79, away from the generation area. However, several restrictions and assumptions should be taken into account when applying these spectra (Sutherland 2010), i.e.:

- There is no mean current and the stratification decreases exponentially with depth from the top of the thermocline and the profile of the buoyancy frequency is $N(z) = N_* \exp(z/h_\sigma)$, where $N_* = 5.2 \times 10^{-3}\text{s}^{-1}$ and $h_\sigma = 1.3\,\text{km}$
- The Coriolis frequency, f, is constant, corresponding to the inertial frequency at 30°N, i.e. $f = 7.3 \times 10^{-5}\text{s}^{-1}$.
- The internal waves are horizontally isotropic and they have sufficiently small amplitude that wavenumber k is related to wave frequency ω by the dispersion relation deduced from linear theory.

The final form of the energy spectrum per unit mass takes the form assuming that waves are separable in vertical wavenumber, k_z, and frequency, ω:

$$F(\omega, j) = E_0 S(\omega) H(j) \tag{6.49}$$

in which $S(\omega)$ is an "universal" fit to the frequency content of motions between f and N:

$$S(\omega) = \frac{2}{\pi}\frac{f}{\omega}\left(\omega^2 - f^2\right)^{-1/2} \tag{6.50}$$

where $f < \omega < N$. The structure of S is defined to reproduce the observed ω^{-2} power law at frequencies $\omega \gg f$. Due to term $\left(\omega^2 - f^2\right)^{-1/2}$, the frequency spectrum S exhibits the observed energy spike for wave close near inertial frequencies.

Function $H(j)$ is an universal fit of the wavenumber content with j being a vertical mode number:

$$H(j) = \left(j^2 + j_*^2\right)^{-1}\left[\sum_{j=1}^{\infty}\left(j^2 + j_*^2\right)^{-1}\right]^{-1}, \quad \sum_{j=1}^{\infty} H(j) = 1 \tag{6.51}$$

where $j_* = 3$ is a cutoff mode number above which the wavenumber shape is red.

The WKB local vertical wavenumber corresponding to mode j is:

$$k_j = \frac{j\pi}{h_\sigma}\left(\frac{N^2 - \omega^2}{N_*^2 - \omega^2}\right)^{1/2} \tag{6.52}$$

Horizontal and vertical spectra are related through the linear dispersion relation:

$$\omega^2 = \frac{N^2 k_h^2 + f^2 k_z^2}{k_h^2 + k_z^2} \tag{6.53}$$

where $k_h^2 = k_x^2 + k_y^2$ is the total horizontal wavenumber.

Functions $S(\omega)$ and $H(j)$ independently describe frequency and wavenumber shape, and E_0 is a constant prescribing the level of the model spectrum. The energy spectrum per unit mass per unit frequency bandwidth and per unit wavenumber bandwidth is:

$$\hat{F}(\omega, j) = h_\sigma^2 N_0 N F(\omega, j) \tag{6.54}$$

The vertical energy spectrum $F(k_z)$ is given by integration of Eq. (6.49) over ω.

The GM spectrum scales as k_z^{-2} for high vertical wavenumber k_z. The Froude number at the dominant vertical wavenumber $k_z \approx 10^{-2}\,\text{m}^{-1}$ is $Uk_z/N \approx 0.1$ with $U \approx 7\,\text{cm s}^{-1}$, which justifies the linear type of wave model.

The total energy per unit mass, integrated over the GM 79 spectrum becomes:

$$E = E_0 h_\sigma^2 N_0 N \left[1 - \frac{2}{\pi} \arcsin\left(\frac{f}{N}\right)\right] \tag{6.55}$$

At present, the Garrett-Munk spectrum serves as a basis for distinguishing between various ocean waters in terms of the mean spectral parameters. It means that this spectrum characterizes some background dynamic state of particular ocean basin. However, in many instances it is the extreme values of wave parameters that are of importance rather than the mean conditions.

References

Bendat JS, Piersol AG (1986) Random data. Analysis and measurement procedures. Wiley, New York, 566 p

Emery WJ, Thomson RE (1997) Data analysis methods in physical oceanography. Pergamon, Amsterdam, 634 p

Garrett C, Munk WH (1975) Space-time scales of internal waves: a progress report. J Geoph Res 80:291–313

Garrett C, Munk WH (1979) Internal waves in the ocean. Annu Rev Fluid Mech 11:339–369

Huang NE, Shen Z, Long SR, Wu MC, Shih HH, Zeng Q, Yen NC, Tung CC, Liu HH (1998) The empirical mode decomposition and the Hilbert spectrum for nonlinear and non-stationary time series analysis. Proc R Soc Lond A454:903–995

Ivanov VA, Pelinovsky EN, Talipova TG (1993) Frequency of internal wave amplitude recurrence in the Mediterranean Sea. Okeanologiya 33(3):180–183 (in Russian)

Massel SR (2001) Wavelet analysis for processing of ocean surface wave records. Ocean Eng 28:957–987

Massel SR (2007) Ocean waves breaking and marine aerosol fluxes. Springer, New York, 323 p

Monin AS, Krasitskii WP (1985) Ocean surface processes. Gidrometeoizdat, Leningrad, 374 p (in Russian)

Munk WH (1981) Internal waves and small scale processes. In: Warren BA, Wunsch C (eds) Evolution of physical oceanography scientific survey in honor of henry stommel. MIT Press, Cambridge, pp 264–291

Pelinovsky EN, Holloway PE, Talipova T (1995) A statistical analysis of extreme events in current variations due to internal waves from the Australian North West Shelf. J Geophys Res 100(C12):24831–24839

Sutherland B (2010) Internal gravity waves. Cambridge University Press, Cambridge, 377 p

Torrence C, Compo GP (1998) A practical guide to wavelet analysis. Bull Am Meteorol Soc 79:61–78

Chapter 7
Internal Waves in the Baltic Sea

7.1 Introduction

In this chapter many theoretical methods developed in the previous chapters are applied to describe the features of internal waves in a shallow sea. The Baltic Sea was chosen for illustration of these methodologies. It is a water basin of special features in terms of the internal dynamics appearance. The most important hydrographical and dynamical characteristic processes and parameters of the Baltic are summarized in Sect. 7.2. In Sect. 7.3, the thermistor chain methodology is described and the experimental data are examined. Section 7.4 is dedicated to the observations of the vertical water structure using CTD sensors moving along the transects in the Southern Baltic Sea (Rak and Wieczorek 2012). All experimental data exhibit the non-regular and random character which requires the application of the spectral methods (see Chap. 6). In shallow waters, establishing the average formula for the internal waves spectrum, valid for a large area is very difficult due to non-uniformity of the sea bottom and possible interaction of the internal wave motion with bathymetry. In fact, in contrast to the deep ocean, the universal spectral distribution of the internal waves does not exist, and resulting formulae are valid only locally.

7.2 Basic Hydrographical and Dynamical Features of the Baltic Sea

The Baltic Sea basin extends about 1300 km north to south (between 54 and 65°N) and about 1100 km east to west (between 10 and 30°). This is a shallow sea with a mean depth of 54 m and the greatest depth of only 459 m in Landsort Deep (Lepparanta and Myrberg 2009). Topography of the sea bottom is very complicated and the water exchange between various sub-basins is largely governed by the sills between them. The only connections with the Atlantic Ocean and North Sea are the shallow Danish Straits. The water is flowing through the Little and the Great Belt over the 18 m

S.R. Massel, *Internal Gravity Waves in the Shallow Seas*,
GeoPlanet: Earth and Planetary Sciences, DOI 10.1007/978-3-319-18908-6_7

deep Dars Sill. The first main basin of the Baltic Sea is the 45 m deep Arkona Basin. From this basin water travels into the Bornholm Basin of max. depth of 95 m. Over the Słupsk Sill (max. depth 60 m) and Słupsk Furrow water flow bifucates further into the Gulf of Gdańsk and into the Gotland Basin. The area of the Bornholm and Gotland Basins is usually denoted as the Baltic Proper. North of the Gotland Basin the water flows east to the Gulf of Finland or north into the Gulf of Bothnia.

Due to very narrow and shallow connection of the Baltic Sea with the Atlantic Ocean, the tides in the Baltic Sea are almost negligible. In fact, the Baltic Sea is one of the largest brackish water area in the world.

However, the saline water masses from the North Sea enter into Baltic Sea from time to time, and with distance from the Kattegat waters become more diluted due to gravitational and intrusive processes, including the internal waves action (Piechura and Beszczyńska-Möller 2004; Feistel et al. 2003, 2004, 2006). The cross-section from south-west to north-east shows that the salinity is changing from 25 $g\,kg^{-1}$, over 9 $g\,kg^{-1}$ in the Arkona Basin and 7 $g\,kg^{-1}$ in the Gotland Basin (Lepparanta and Myrberg 2009). The salinity varies little over the year, but it is strongly depth-dependent, changing from 7.5 $g\,kg^{-1}$ at the surface to 17 $g\,kg^{-1}$ at the bottom. The HELCOM BMPK02 station's long observation series shows strong seasonal dependence of water temperature, especially in the surface layer, where temperature changes from 2.5 °C at the end of winter till 17.9 °C in August.

Barotropic inflows of saline water from the North Sea are caused by the specific meteorological conditions, when initially prolonged easterly winds push the water out of the Baltic Sea into the North Sea. After that, if the wind direction changes, or the wind speeds drop, water flows back into the Baltic Sea, establishing the pressure equilibrium again. A particular role is played by the rapid inflows which transport hundreds of cubic kilometers of well oxygenated, saline waters into Baltic Sea within a few days (Piechura and Beszczyńska-Möller 2004). According Reissmann et al. (2009), the typical inflow of 200 km^3 of water imports about 2×10^9 tonnes of salt and 1×10^6 tonnes of oxygen.

Exchange of saline water between Bornholm and downstream Gdańsk and Gotland Basins occurs through the Słupsk Furrow which is separated from the upstream basin by a shallow 60 m deep sill (Fig. 7.1). When enough dense deep water is accumulated in the Bornholm Basin and this water reaches the level of the Słupsk Sill, it overflows and continues to move through the Słupsk Furrow into the Gdańsk and Gotland Basins. When the water climbs over the sill and flows down along the sloping bottom, the volume of gravity current grows due to entrainment and mixing.

On the other hand, during prolonged periods of calm weather and with strongly stratified Belt Sea, baroclinic inflows may occur (Feistel et al. 2003, 2004). They are driven by baroclinic pressure gradients. During a baroclinic inflow the inflowing water is salty, but also warm, when compared to the cold water flowing during the barotropic inflows in winter.

Vertical structure of water masses exhibits a pronounced salinity gradient with depth of halocline depending on the depth of the basin, with 50 m in the Bornholm Basin and 65–70 m in the Gotland Basin. The thermocline is subject to seasonal variation, from shallower in the summer time and deeper in winter.

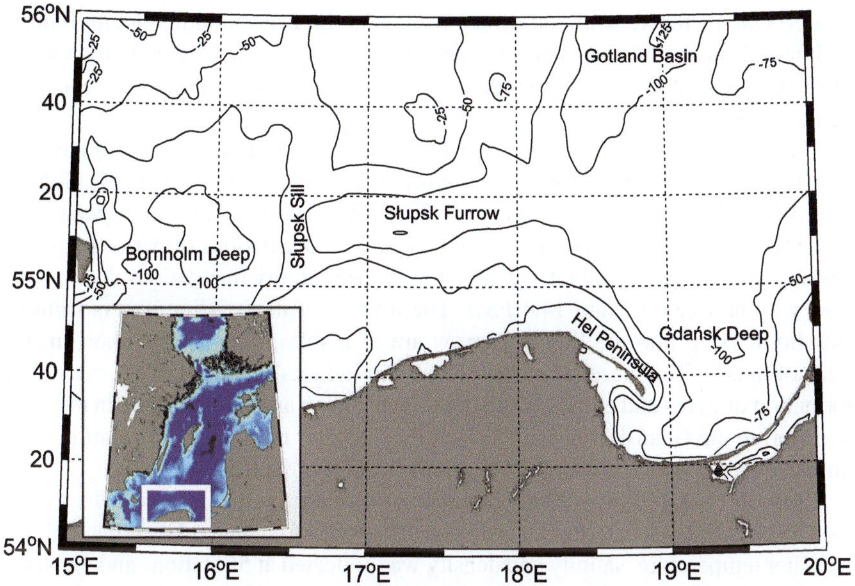

Fig. 7.1 Bornholm Basin, Słupsk Furrow and Gdańsk Deep

A common feature in the Baltic Sea are mesoscale eddies, which form circulating lenses of water. Eddies may contain water of properties different from than the surrounding water. Within the core of rotating eddies, the water and isopycnals are moving up or down depending on the temperature. The warm-core eddies rotate anticyclonically and isotherms are displaced downward. In the Baltic Sea, the mesoscale eddies have been given a nickname: they are called Beddies. Usually the Beddies have radii of 4–30 km and temperature anomalies of e.g. 4 °C (Reissmann et al. 2009). Due to lack of barotropic tidal pressure gradients, internal tides are not relevant energy sources for diapycnal mixing in the bottom boundary layer. Therefore, in terms of mixing, the Baltic Sea becomes similar to lakes or fjords with weak tidal forcing. In such environments, the long internal waves and internal seiches play the role of the internal tides. In terms of the spectral energy distribution, the short internal waves, inertial-internal waves and coastly trapped long internal waves are the main components.

Baltic Sea is also subject to strong wind forcing with time scales of a few days. The response of the sea to wind and buoyancy forcing shows a high degree of mesoscale patterns in the Baltic, ranging typically from 5 to 20 km. The strong winds of 10–15 m s^{-1} generate internal waves with amplitudes of 11–15 m and currents velocities of 5–8 cm s^{-1} in the bottom layers (Chernysheva 1987).

Some data on the internal waves characteristics observed in the Baltic Sea were published by Kol'chinskii et al. (1996), Golenko and Mel'nikov (2007), and others. Kurkina et al. (2011) examined the geographical and seasonal distributions of kinematic parameters of long internal waves using the Generalised Digital Environmental

Model. One of the very important parameters is the near-bottom velocity induced by internal waves for cases when the density jump layers are located in the vicinity of seabed. Also flow over topography can generate near-inertial waves (Fennel and Schmidt 1993).

There is a considerable amount of data on the structure and dynamics of the water flow over the Słupsk Sill (Kõuts and Omstedt 1993; Piechura et al. 1997; Golenko et al. 1999; Feistel et al. 2006; Paka et al. 2006 ; Zhurbas et al. 2011). The water flow in the Słupsk Furrow area is very complicated with oscillations of periods covering a wide range, from two minutes to about 30 h. The salinity fluctuations due to water flow vary between 9.2 and 14.4 PSU. The inflows of more saline waters from the Bornholm Deep into the Słupsk Furrow are of a splash-like nature, short in time (tens of hours). The presence of the Słupsk Sill induces internal wave breaking and formation of gyres and eddies. These phenomena play an important role in the water transport to the Gdańsk Bay and Gotland Basin, water mixing, and formation of the intrusive lenses which are a result of intensification of internal waves.

Talipova et al. (1998) studied the kinematic characteristics of internal waves in the Gotland Basin where the intrusive lenses appear frequently. Experimental data on water temperature, salinity and density was collected at 56 stations and the distribution of phase speed C_0, the nonlinearity coefficients α and dispersion coefficients β of the KdV equation were determined. They all appear to be mostly functions of water depth and vertical density structure.

The studies on the kinematic characteristics of internal waves in the Baltic Sea were extended on the World Ocean by Pelinovsky et al. (2000). The initial seasonal profiles of temperature and salinity were based on the World Ocean Atlas 1994 (Levitus 1994). Using the world seasonal data, the vertical profiles of the Väisäla-Brunt frequency was determined and the internal wave phase speeds were presented on the maps of one degree resolution. Geographical and seasonal variability of phase speed have been calculated. It was shown that seasonal variation of the phase speed is less than 10 % in most of the basins of the World Ocean.

In an almost closed sea basin, such as the Baltic Sea, longer oscillations are induced by seiches (Lepparanta and Myrberg 2009). Seiches play an important role for the Baltic Sea, causing very high sea level when favourably coupled with wind-driven sea level variations. Their period depends on the size and depth of the basin. Two-dimensional models showed that oscillation modes of the whole Baltic Sea vary from 31 h for the first mode to 7.3 h for mode $n = 10$, when the Coriolis effect is taken into account. On the other hand, observations indicate the existence of 17 and 19 h oscillations periods. The system Baltic Sea Proper—Gulf of Bothnia has periods as long as 39 h. Similar oscillation periods of 24–36 h for the whole Baltic Sea are resulting from the simple Merian formula (Lepparanta and Myrberg 2009).

For the small basins the periods are shorter, of order of some hours. For example, for the Gulf of Gdańsk, the seiches are of period $T = 3.5$–4 h, while the period of transverse oscillations between southern coast of Baltic and southern coast of Scandinavia is about 5.5–7.5 h. On the other hand, frequencies higher than the buoyancy frequency do not represent freely propagating internal waves, but evanescent waves, instabilities or turbulent scale phenomena. Internal waves with a frequency near N have a low-modal structure and occur in packets. For a shallow sea, as the Baltic Sea, the motions in the upper part of the stratified water column are directly affected by processes in the surface layer.

Kol'chinskii et al. (1996) developed an approximate theoretical model for seiche oscillations in the Słupsk Furrow treated as the rectangular basin with constant water depth. They found that the ratio of the oscillation period T_n of the mode, n, to the inertial wave period T_f is:

$$\frac{T_n}{T_f} = \left(1 + \left(\frac{kh}{n\pi}\right)^2\right)^{1/2} \left(1 + \left(\frac{kh}{n\pi}\right)^2 \left(\frac{T_f}{T_N}\right)^2\right)^{-1/2} \tag{7.1}$$

in which h is the water depth, $k \approx n\pi/M$, $T_N = 2\pi/N$. Assuming depth $h = 100$ m, width od the Słupsk Furrow $M = 30$ km, period associated with frequency N, $T_N = 6$ min and inertial period $T_f = 15$ h, the seiche period of the first mode becomes equal 13.5 h.

7.3 Observations of Internal Waves Using Thermistors Chain

Internal waves passing a given point in the water column generate displacement of the isotherms from their initial position. As the temperature of water particle can be treated as a passive indicator following the water movement, records the temperature fluctuations are interpreted as internal waves presence. However, only undulations of water density, ρ, are closely related to the internal waves motion and displacement of the isopycnals give us correct information on the dynamics of internal waves. Nevertheless, recording of isotherms provides very useful information on the internal waves signal at a given location and time. Therefore, a moored thermistor chain is a popular method to recording rising and falling of the temperature at a given thermistor sensor level, even in the Arctic conditions (Thorpe 2005; Morozov and Marchenko 2012).

In Fig. 7.2, temperature variations at four selected thermistor levels are shown. The thermistor chain was moored at the Station Hel ($\lambda = 18°52.132$, $\varphi = 54°39.633$) in the Southern Baltic in April 2014. Thermistor sensors span the range from 51 to 70 m below the surface in a water depth of 76 m at the interval of 1 m. Temperature was sampled with $\Delta t = 10$ s. It can be seen that the temperature at the selected level is non-stationary and slowly rising in time. For the given observation period, the water temperature at the bottom is higher than the temperature at the levels above due to inflow of more hot waters from the Bornholm Basin.

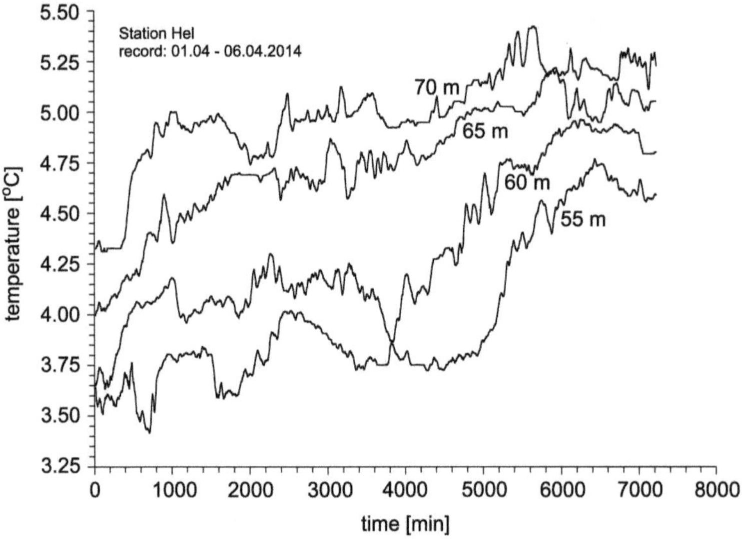

Fig. 7.2 Temperature variations at four selected water depths

To get some impression of the internal waves dynamics and to determine their parameters it is more useful to consider the displacement of the particular isotherm in time. The time series of isotherm $T = 4.4\,^{\circ}\mathrm{C}$ of duration of about 70 h was extracted from the records of thermistor chain. To determine the Fourier Transform, removal of the spurious trend is usually required. In our case, filtering of the experimental data prior to a detailed analysis is needed for removing the isotherm displacement variations of the seasonal scale and possible changes due to influx of water from different Baltic Sea basins to the Peninsula Hel region. After the trend removal we get a record of the isotherm displacements as shown in Fig. 7.3. Irregular internal waves pattern with amplitude of about 2.5 m is clearly seen. These larger waves are superimposed by shorter oscillations of smaller amplitude.

The corresponding spectral energy density is shown in Fig. 7.4. The energy cascades down to smaller scales, through a frequency continuum till total dissipation. Three higher peaks are concentrated in the low frequency range. It should be noted, however, that the temperature induced by the internal waves is restricted only to the frequencies greater than inertial frequency f. In our case, inertial frequency $f = 2\Omega \sin\varphi = 0.001136\,(1/\mathrm{min})$, which corresponds to the period $T \approx 880\,\mathrm{min}$ (14.67 h). Therefore, the highest peak in Fig. 7.4 in the spectral density of temperature variation is due to oscillations longer than the inertial waves. Within the internal frequency range, the highest peaks have periods of 11.9 and 7.57 h.

The Fourier Transform yields information on how much, but not when (in time), the particular frequency components exist. Such information is sufficient in a case of the stationary signals as the frequency content of such signals does not change in time and all frequency components exist all the time. The isotherm evolution in Fig. 7.3

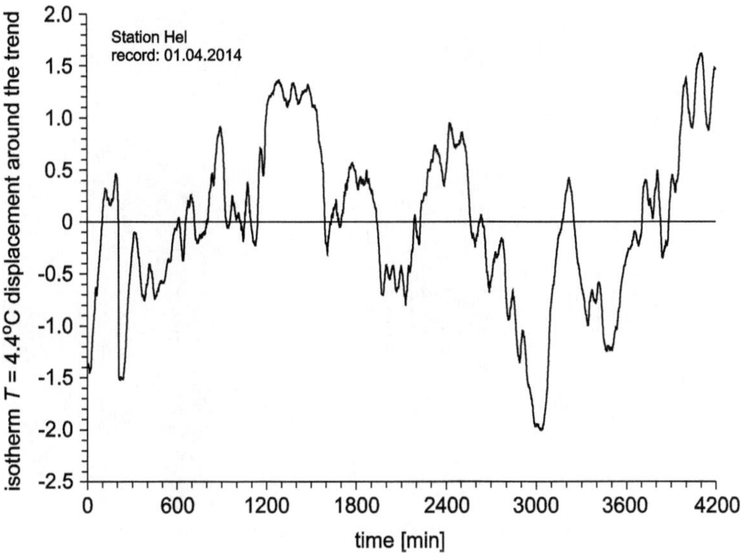

Fig. 7.3 Displacements of the isotherm $T = 4.4\,°\mathrm{C}$ after trend removal

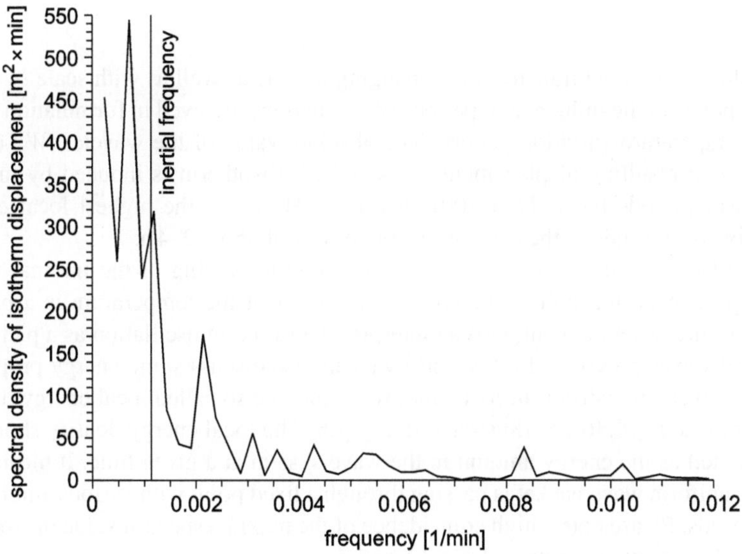

Fig. 7.4 Spectral density for isotherm ($T = 4.4\,°\mathrm{C}$) displacements. Corresponding time series is given in Fig. 7.3

is not strictly stationary. Therefore it will be useful to examine this time series by applying the Wavelet Transform discussed in Chap. 6 for the displacements of the isotherm $T = 4.4\,°\mathrm{C}$ recorded at the Station Hel on April 1, 2014. Figure 7.5 shows the contours of the absolute value of the wavelet transform $|WT|$ (see Eq. 6.28).

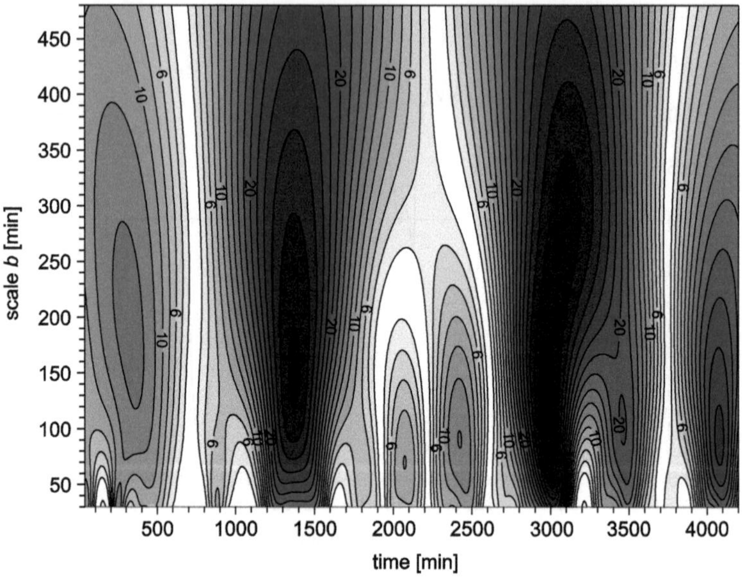

Fig. 7.5 Contours of the absolute value of wavelet transform $|WT|$

Values of wavelet transform are changing in time as well as with scale b, which corresponds to the influence of periods of oscillations involved in formulation of the final temperature variation pattern. The absolute value of the wavelet $|WT(\tau, b)|$ shows that resulting displacement of the selected isotherm is induced by internal waves of periods from about 100 min till 8 h. However, the highest local energy density corresponds to the oscillations of periods of about 2–4 h.

Displacement of the isotherms in time is due to passing of the internal waves through a recording point. Therefore, assuming that the temperature is a passive index of the water element, we can interpret the isotherm oscillation as a picture of internal waves presence. In this sense we can also discuss some energy properties of the wavelet transform. In particular, we define the so called local energy density $CE_2(\tau)$ (see Eq. 6.36), as illustrated in Fig. 7.6. The local energy density should be interpreted as the energy amount in the wave system at a given time. It means that a-non-uniform wave packet is passing through a fixed point with various amplitudes and speeds. Please note a high coincidence of the maxima spectral value of isotherm $T = 4.4\,°\text{C}$ displacements on Figs. 7.5 and 7.6.

Let us now represent the time series of isotherm displacements of 70 h duration in the form of the Fourier series according to Eq. (6.40) and develop a complex analytical signal η (see Eq. 6.42) applying the Hilbert Transform. Therefore, the real part of the expression $A(t)\exp[i\phi(t)]$ provides a very compact representation of initial signal

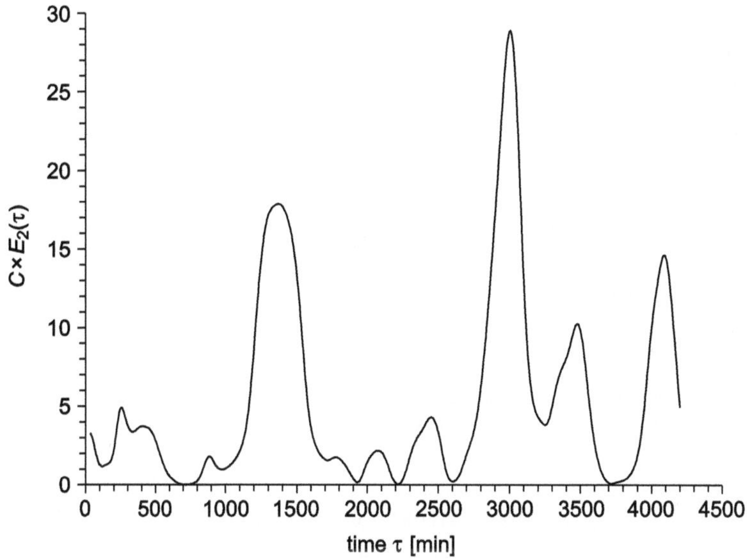

Fig. 7.6 Local energy density $CE_2(\tau)$

$\zeta(t)$. It was shown that this expression with $n = 1800$ terms in the Fourier series (see Eq. 6.40) yields isotherm displacements, which are totally non-distinguishing from the initial displacements $\zeta(t)$. Very valuable feature of the Hilbert Transform is a possibility of determination of the kinematics of internal waves from the initial time series $\zeta(t)$ in a simple way. This information is of special importance for marine ecologists as the internal waves are one of the basic mechanisms for transportation of nutrients, marine organisms and pollutants in the water column, particularly in the vertical direction. Velocities of isotherm displacements result from Eqs. (6.47) and (6.48). The instantaneous vertical velocity $w(t)$ and horizontal velocity $u(t)$ are shown in Figs. 7.7 and 7.8. The initial isotherm displacements are given for comparison.

As was mentioned above, the variation of the isotherms in time can serve as an indicator of the presence of internal waves. However, to determine the internal waves characteristics we have to go into more detail and consider displacements of the selected isopycnal in time depending on the instantaneous values of temperature, T, and salinity, S. Therefore, let us assume that temperature is known from the thermistor chain records described above. The CTD probe installed at the Station Hel showed that salinity was almost unchanged during about 70 h of recording. In vertical direction, salinity varies approximately linearly from $\approx$6.2 PSU at the sea surface to $\approx$10.66 PSU at sea bottom. Combining values of recorded temperature, $T(t, z)$, and interpolated salinity values, $S(t, z)$, we are able to determine the density $\rho(t, z)$ for the water depth from 51 to 70 m below sea surface, where the thermistor sensors were installed. The water density follows from the UNESCO Formula (1981) described in Appendix A. In Fig. 7.9, the resulting time series of the isopycnals

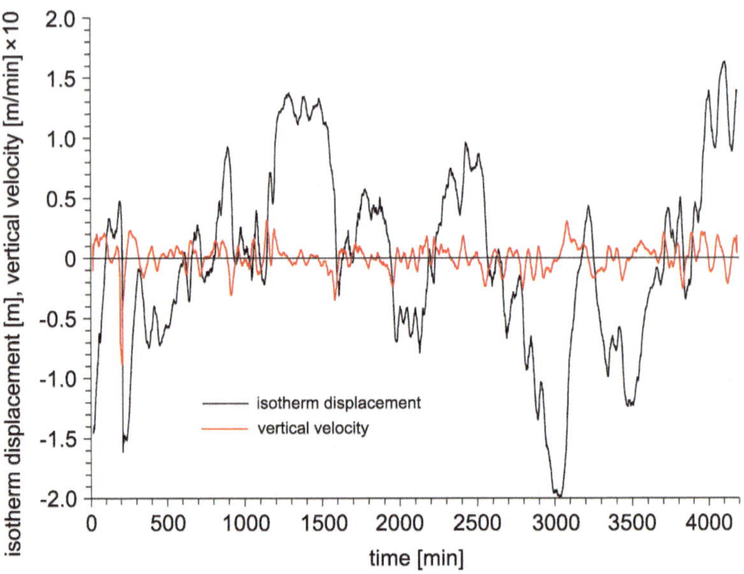

Fig. 7.7 Vertical velocity corresponding to isotherm displacements $\zeta(t)$

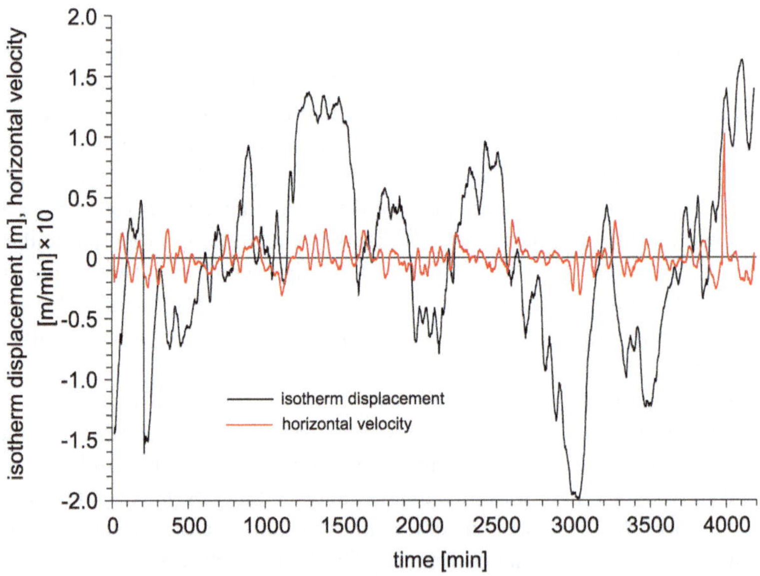

Fig. 7.8 Horizontal velocity corresponding to isotherm displacements $\zeta(t)$

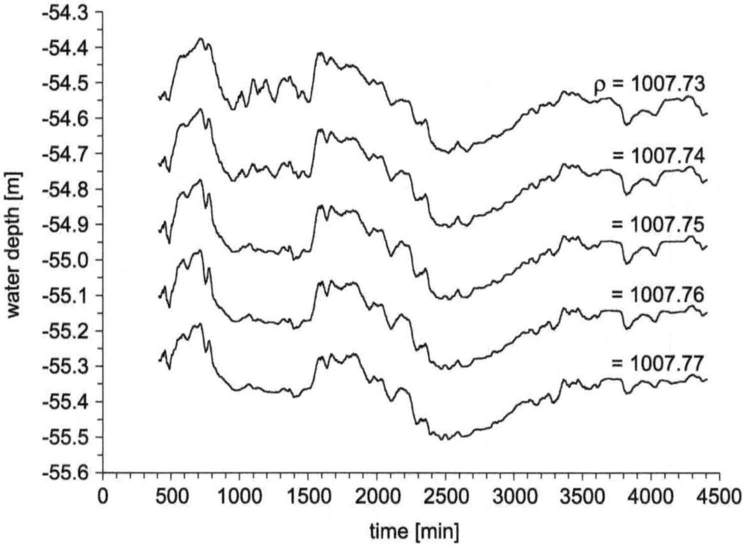

Fig. 7.9 Time series of isopycnals $1007.73 < \rho < 1007.77$ displacements

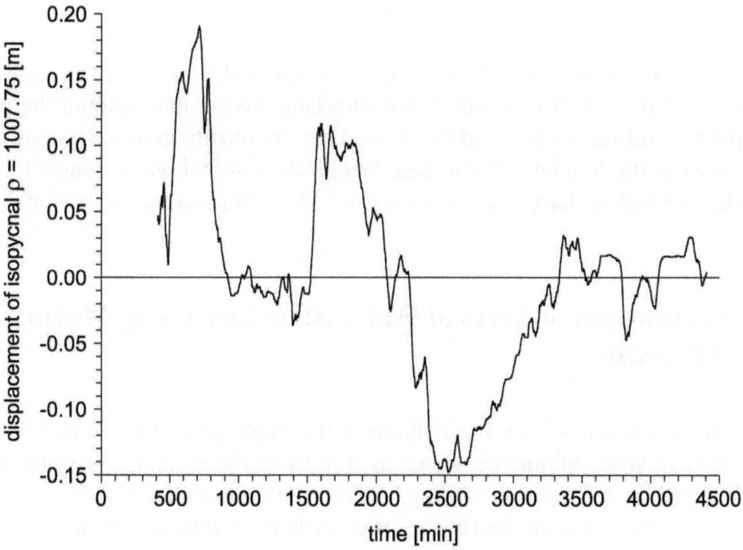

Fig. 7.10 Time series of isopycnal $\rho = 1007.75$ displacement

$1007.73 < \rho < 1007.77$ are given for illustration. A more detailed time series of selected isopycnal $\rho = 1007.75\,\mathrm{kgm}^{-3}$ is shown in Fig. 7.10. Very long oscillations of isopycnals are supplemented by very short waves of small amplitudes. During record time of $\approx$67 h, the isopycnals oscillated between $+0.2$ and -0.15 m against

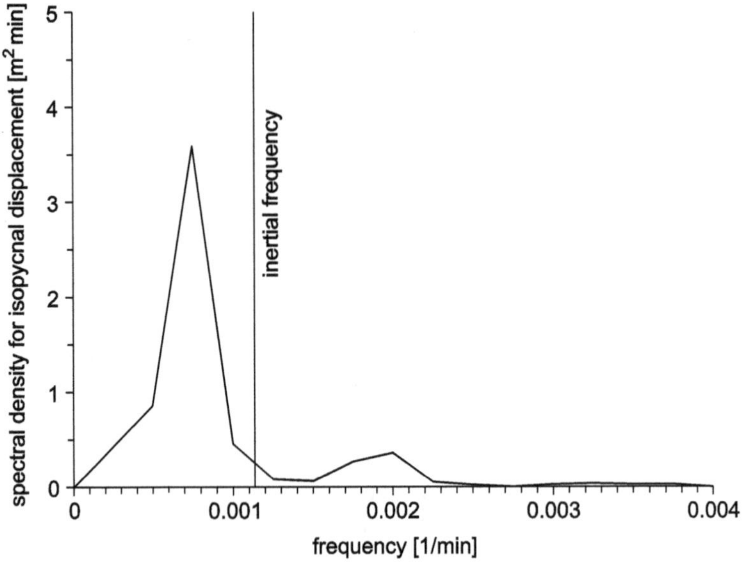

Fig. 7.11 Spectral density for isopycnal $\rho = 1007.75$ displacement

the mean reference water depth of $\approx$55 m. The spectral density for this time series is shown in Fig. 7.11. This result is not surprising taking into account the pattern of isopycnal oscillations given in Fig. 7.10. They are mostly induced by long seiche oscillations in the Southern Baltic Sea. Within the internal waves range, the most energetic oscillations have periods from 7 to 11 h, with maximum at period 8.3 h.

7.4 Observations of Vertical Water Structure Using Moving CTD Sensors

The vertical structure of a water column in the Southern Baltic strongly depends on the time of year, summer or winter, as well as on the exchange of water masses with the North Sea. Significant regional and seasonal variations of the Väisälä-Brunt frequency profiles are controlled by the different thermohaline stratification and water depth in different basins. The strongest seasonal variations occur in the Belt Sea and in the Arkona Basin (Fennel et al. 1991). For Bornholm Basin and Baltic Proper, the yearly cycle of the Väisälä-Brunt frequency is mostly controlled by the development of a thermocline in summer while the halocline remains almost unchanged.

In the following we consider the winter hydrological scenario of the year 2003. In that time, a medium-size inflow of about 200 km^3 of saline water from the North Sea occurred during January–February. This inflow brought very cold water, of temperatures around 1–2 °C and less. Since the temperature of deep water in the Southern

Baltic before the inflow was exceptionally high, the inflowing water induced large changes in the water structure. Inflowing waters subsequently migrated through the deep basins and channels from Arkona Basin to the Gdańsk Deep during next 4–8 months. The strong mesoscale structures and intensive mixing followed the eastward transport of the inflow water, particularly in the Bornholm Deep and Słupsk Furrow. Details of the inflow waters in the spring 2003 can be found in the papers by Piechura and Beszczyńska-Möller (2004) and Feistel et al. (2003).

For the analysis, we use the temperature and salinity data measured during the cruise of the Polish Academy of Sciences research vessel *Oceania* with towed CTD twice in February 2003 along the transect in the Southern Baltic Sea, from Bornholm Basin, through Słupsk Sill and Słupsk Furrow to Gdańsk Basin, along the latitude $\varphi \approx 55.25°$ (see Fig. 7.1). The temperature and salinity pattern recorded on 4–7 February 2003 is given in Fig. 7.12.

In the Słupsk Furrow, a sill blocks the flow substantially as the water depth at the Słupsk Sill is only ≈56 m while the water depths upstream and downstream of the sill are of the order of 90–100 m. The western upstream slope of the Słupsk Sill is of the order of 1.5×10^{-3}, while the downstream slope is of the order 7.5×10^{-4}. During water inflow and intensive mixing, a large body of saline water is climbing on the eastern slope of the Bornholm Deep (Piechura and Beszczyńska-Möller 2004). At a top of the Słupsk Sill water with temperature of 4–6 °C and salinity of 13–14 PSU in the bottom layer is observed. In the Słupsk Furrow the warm waters are pushed eastward and upwards, and the deep water layer is extended upwards while the salinity is increased by about 1 PSU.

To characterise the vertical water structure, profiles of water temperature T, salinity S and potential density σ_t at stations along the transect have been determined. Examples of the T, S and σ_t profiles for two stations are given in Figs. 7.13 and 7.14. At both stations, a three-level density structure is clearly seen. In the upper layer, water is well mixed and vertical profiles of temperature, salinity and density are uniform. Also, the bottom layer is characterised by almost constant values of these parameters. Between upper and bottom layers, the pycnocline with bigger gradients of temperature, salinity and density is located. When water depth becomes smaller at the Słupsk Sill, the thickness of the bottom layer is strongly reduced.

In Table 7.1, the thicknesses of the particular layers and some non-dimensional characteristic parameters of the pycnocline for the transect of 4–6 February 2003 are given. The distance (in km) for particular stations is measured from the Słupsk Sill with positive direction toward the east. The depth h_1 denotes water depth at the top of the pycnocline, h_2 is the depth at the pycnocline bottom, while h is the total local water depth. Non-dimensional thickness of the pycnocline $(h_2 - h_1)/h$ varies from ~0.08 to ~0.38, depending on the local water depth. At the Słupsk Sill, bottom layer is particularly small, and upper layer of the well mixed water extends up to 99 % of local water depth.

The methodology of determination of the vertical water structure and isopycnal variation is described in Appendix B. It was found that isopycnal submergence h_ρ is a function of distance x and time t, i.e. $h_\rho = f(x, t)$. Due to approximately constant ship speed, variation of this submergence lies in the cross-section of the space $h_\rho =$

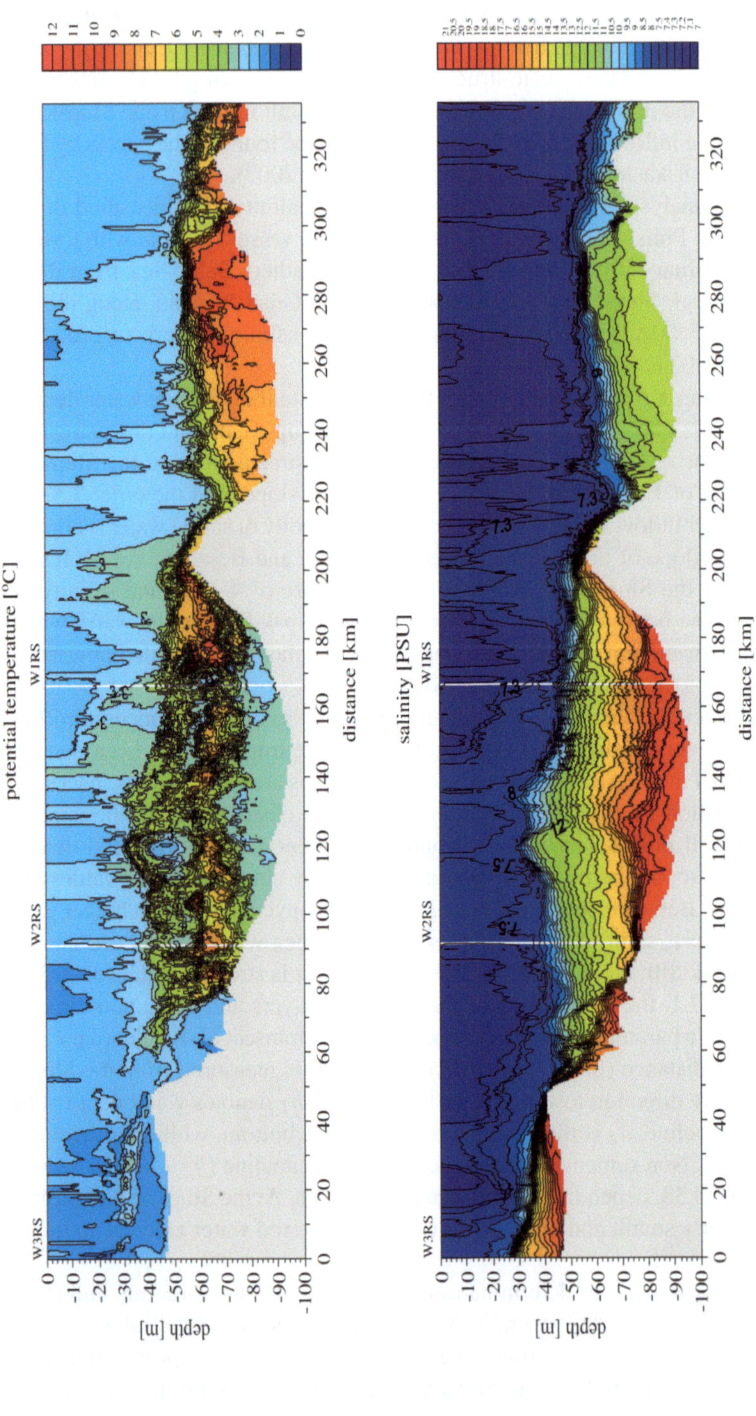

Fig. 7.12 Temperature and salinity pattern along transect in the Southern Baltic Sea (Piechura and Beszczyńska-Möller 2004)

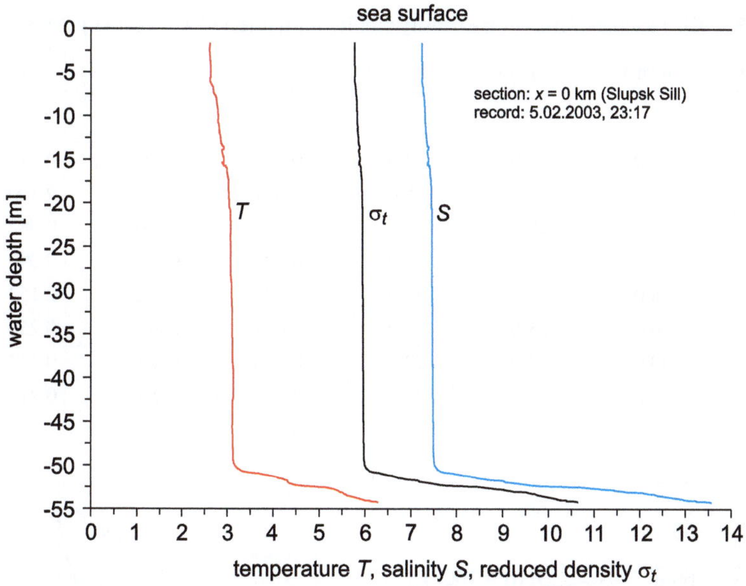

Fig. 7.13 Vertical profiles of water temperature T, salinity S and density σ_t at the Słupsk Sill

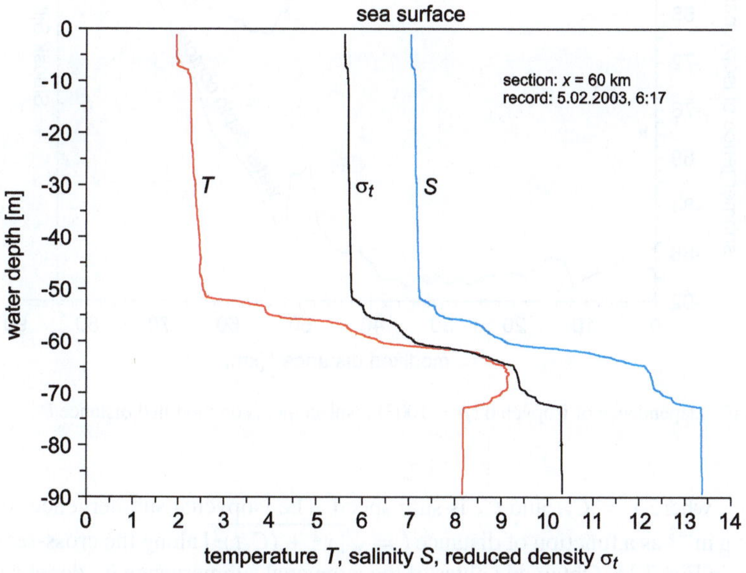

Fig. 7.14 Vertical profiles of water temperature T, salinity S and density σ_t at station of 60 km east of the Słupsk Sill

Table 7.1 Characteristic levels of the density profiles for transect of 4–6 February 2003

Location of stations (km)	h (m)	h_1 (m)	h_2 (m)	$\frac{h_1}{h}$	$\frac{h_2}{h}$	$\frac{h_2 - h_1}{h}$
−20	85.0	46.0	78.0	0.541	0.918	0.377
−10	66.5	45.0	65.0	0.677	0.977	0.300
0	55.0	50.0	54.5	0.909	0.991	0.082
5	67.0	60.0	66.0	0.895	0.985	0.090
15	85.0	62.0	75.0	0.729	0.882	0.153
35	90.0	55.0	78.0	0.611	0.867	0.256
55	90.0	52.0	72.0	0.578	0.800	0.222
75	80.0	55.0	72.0	0.687	0.900	0.213

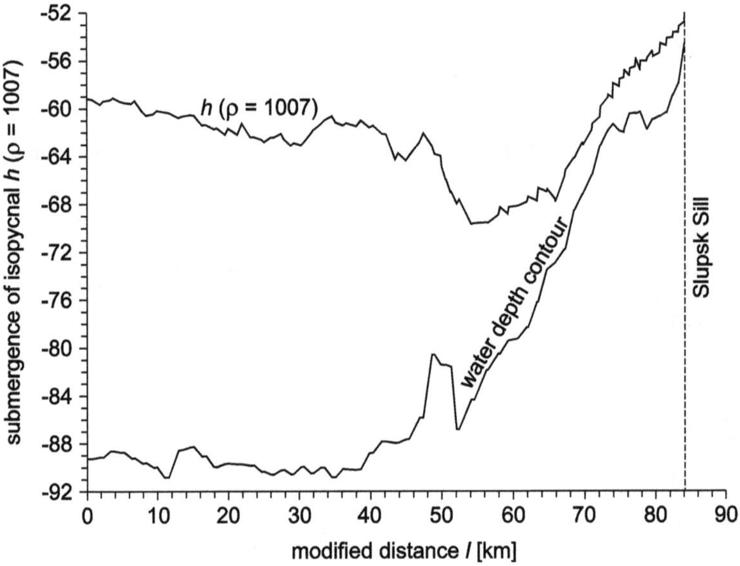

Fig. 7.15 Dependence of isopycnal ($\rho = 1007$) displacements on modified distance l

$f(x, t)$, where $x = C_s t$ and C_s is ship speed. The isopycnal submergence for $\rho =$ 1007 kg m^{-3} as a function of distance $l = \sqrt{[x^2 + (C_s t)^2]}$ along the cross-section is shown in Fig. 7.15. Obtained values of the isopycnal submergence h_ρ depend on the local water depth and the ratio h_ρ/h is changing from 0.66 till 0.98 at the Słupsk Sill. For a moving sensor, when $x = C_s t$, the function $h_\rho = f(x)$ should be identical to the function $h_\rho = f(t)$, after scaling $t = x/C_s$. Comparison of both functions is shown in Fig. 7.16 for isopycnal $\rho = 1007$ kg m^{-3}. However, in real case, the ship speed is not exactly constant and with increasing time of observation, the discrepancy between functions $h_{\rho=1007} = f(x)$ and $h_{\rho=1007} = f(t)$ is going to be bigger. Also,

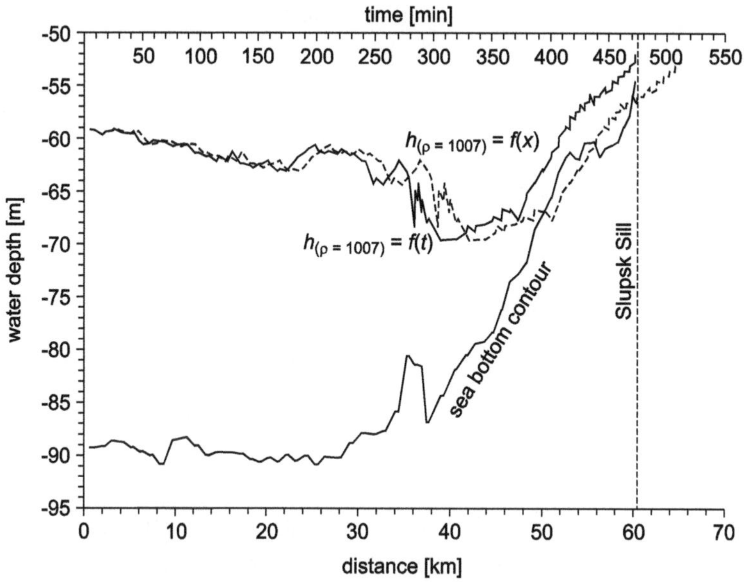

Fig. 7.16 Dependence of isopycnal ($\rho = 1007$) submergence on distance and time

the isopycnal depth varies in space due to changing of water depth, especially on the eastern slope of the Słupsk Sill.

Cruises of r/v *Oceania* in August 2003–May 2004 showed the propagation and transformation of the warm water layer in the Słupsk Furrow and Gdańsk Basin (Feistel et al. 2004). In January 2004, the warm layer in general moved to the east, which was recorded in the Słupsk Furrow and on the slope towards the Gdańsk Basin. We explored the isopycnals variations for the eastern part of the 2004 year transect, starting from the Gdańsk Deep and following westward. Two characteristic vertical profiles of the temperature, salinity and reduced density are shown in Figs. 7.17 and 7.18. The three-layer water column structure is still seen and thickness of the intermediate layer becomes smaller when water depth is reduced from 110 m to about 82 m.

In Fig. 7.19 the variation of the three isopycnals in distance from the Gdańsk Deep towards the west is shown. At greater water depth, the internal waves are clearly exhibited. Wave height is about 8 m for isopycnal $\rho = 1007\,\text{kg}\,\text{m}^{-3}$ and 12 m for isopycnal $\rho = 1009\,\text{kg}\,\text{m}^{-3}$. Corresponding wave length is very large, about 20 km.

For proper interpretation of the experimental results, an influence of the ship motion and resulting Doppler effect should be considered. Doppler effect depends on relation between the ship speed and the phase speed of the internal waves. For the r/v *Oceania* cruises, mentioned above, the ship speed was about $2\,\text{m}\,\text{s}^{-3}$. However, the phase speed C_0 should be determined for a given vertical structure of water masses from Eq. (3.78). The vertical eigenvalues λ are completely determined by the

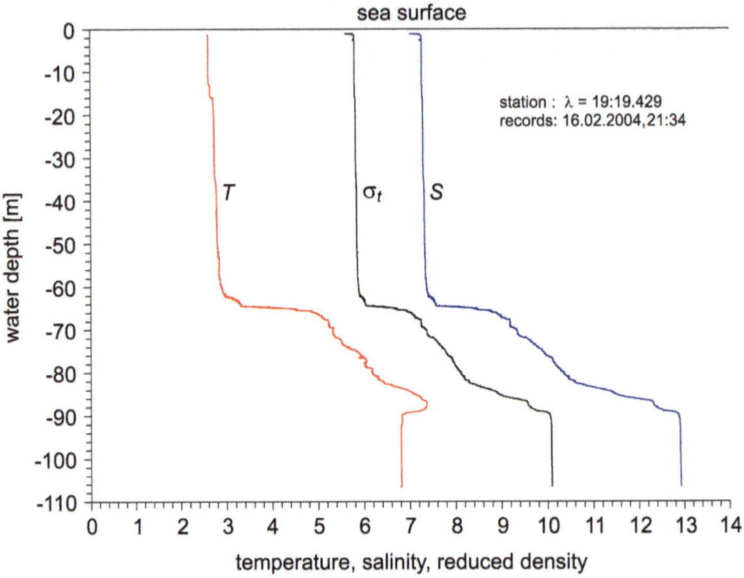

Fig. 7.17 Vertical profiles of water temperature T, salinity S and density σ_t in the Gdańsk Deep

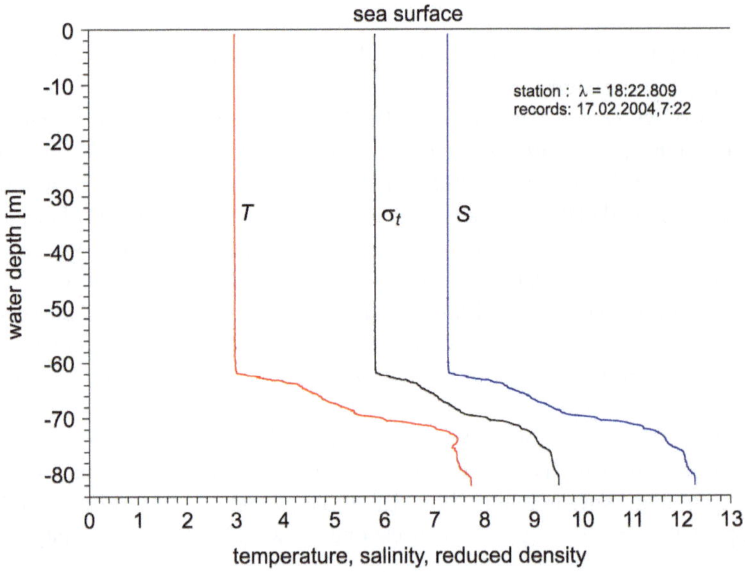

Fig. 7.18 Vertical profiles of water temperature T, salinity S and density σ_t at 60 km west from Gdańsk Deep

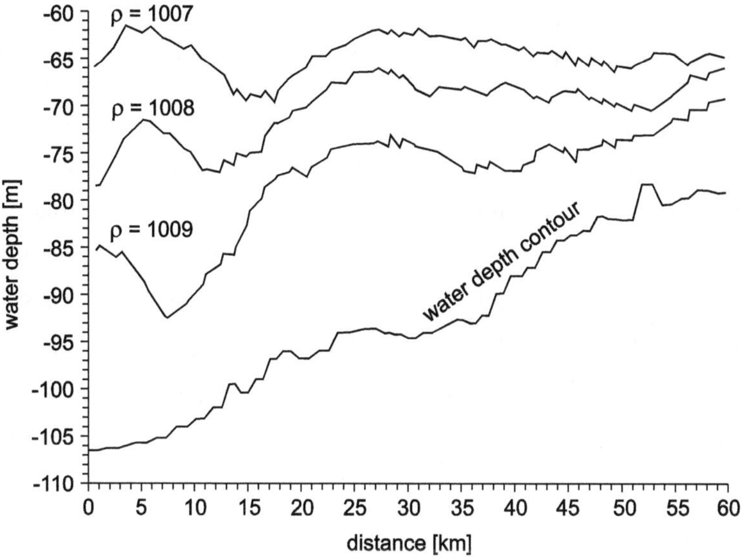

Fig. 7.19 Isopycnal contours as a function of distance from Gdańsk Deep

vertical profile of the Väisälä-Brunt frequency $N(z)$ and water depth h. The phase speed $C_0^{(n)}$ appears as an inverse eigenvalue, i.e. $C_0^{(n)} = 1/\sqrt{\lambda}$. In Table 7.2, the phase velocities $C_0^{(n)}$ are shown for some stations in the Southern Baltic, located at the constant latitude $\varphi = 55.25°$N. It should be noted that the phase speed for the first mode $C_0^{(1)}$ varies between 0.3 and 0.8 m s^{-1} and it is much lower then the ship speed. The phase velocities summarized in Table 7.2 are similar to the velocities given by Talipova et al. (1998) for the Gotland Deep.

Due to high variability of the mesoscale pattern in the Baltic Sea, the internal Rossby radius of deformation appears to be an important quantity in the determination of the horizontal scales and influence of the inertial frequency f for the internal wave modelling. The Rossby radius is defined as:

$$R_1^{(n)} = \frac{C_0^{(n)}}{f} \tag{7.2}$$

in which $C_0^{(n)}$ is the internal wave phase velocity of the n-mode and f is the inertial frequency. In particular:

$$R_1^{(1)} = \frac{C_0^{(1)}}{f} \tag{7.3}$$

is the baroclinic or internal Rossby radius, while barotropic or external Rossby radius is:

Table 7.2 Phase speed and Rossby radii

	Station 190 λ=16°19.180	Station 220 λ=16°49.354	Station 240 λ=17°6.697	Station 260 λ=17°25.623	Station 280 λ=17°46.313
h (m)	65.0	84.0	90.0	89.0	78.0
$C_0^{(1)}$ (m s^{-1})	0.380	0.278	0.500	0.801	0.360
$C_0^{(2)}$ (m s^{-1})	0.060	No data	0.100	0.216	0.060
$C_0^{(3)}$ (m s^{-1})	No data	No data	No data	0.137	0.040
$R_1^{(1)}$ (km)	3.173	2.321	4.173	6.686	3.005
$R_1^{(2)}$ (km)	0.501	No data	0.845	1.803	0.501
$R_1^{(3)}$ (km)	No data	No data	No data	1.144	0.333
R_0 (km)	220.822	239.663	247.984	246.603	230.889

$$R_0 = \frac{\sqrt{gh}}{f} \tag{7.4}$$

The barotropic Rossby radius is very large, comparable to the basin scale of the Baltic Sea. As the barotropic phase velocity $C_0 = \sqrt{gh}$ is high, the barotropic processes are communicated very quickly through the whole sea basin.

From Table 7.2 it follows that internal Rossby radius $R_1^{(1)}$ changes from 2.3 till 6.6 km. It should be noted that these values have been obtained for the winter stratification and they are in agreement with values reported by Fennel et al. (1991).

For the Baltic Proper, $C_0^{(1)} \approx 0.61\,\mathrm{m\,s^{-1}}$ and $R_1^{(1)} = 5.1\,\mathrm{km}$. These values are smaller for the shallower Arkona Basin, i.e. $C_0^{(1)} \approx 0.27\,\mathrm{m\,s^{-1}}$ and $R_1^{(1)} \approx 2.27\,\mathrm{km}$. In other seasons, the phase speed $C_0^{(n)}$ and Rossby radii $R_1^{(n)}$ vary a little.

References

Chernysheva ES (1987) On the modelling of long internal waves. In: Davidan IE (ed) Problems of modelling of ecosystem of the Baltic Sea. Gidrometeoizdat, 50–54 (in Russian)

Feistel R, Nausch G, Mohrholz V, Lysiak-Pastuszak E, Seifert T, Matthaüs W, Krüger S, Hansen IS (2003) Warm waters of summer 2002 in the deep Baltic proper. Oceanologia 45(4):571–592

Feistel R, Nausch G, Heene T, Piechura J, Hagen E (2004) Evidence for a warm water Infow into the Baltic proper in summer 2003. Oceanologia 46(4):581–598

Feistel R, Nausch G, Hagen E (2006) Unusual Baltic inflow activity in 2002–2003 and varying deep-water properties. Oceanologia 48(S):21–35

Fennel W, Schmidt M (1993) Topographically forced inertial waves. J Mar Syst 4(1):1–15

Fennel W, Seifert T, Kayser B (1991) Rossby radii and phase speeds in the Baltic Sea. Cont Shelf Res 11(1):23–36

Golenko NN, Beszczyńska-Möller A, Piechura J, Walczowski W (1999) Some results of research on internal waves in the Stolpe Sill area. Oceanologia 41:537–551

Golenko NN, Mel'nikov VA (2007) Estimation of spatio-temporal parameters of the internal wave field in the southwest Baltic Sea using data by towed probe. Scientific Rep. Russian Geographic Union, Russian State Univ, Kaliningrad 5:C1–C4 (in Russian)

Kol'chinskii NN, Monin AS, Paka VT (1996) On internal seiches in the deep Baltic Sea. Doklady Akademii Nauk 346(2):249–255 (in Russian)

Kõuts T, Omstedt A (1993) Deep water exchange in the Baltic proper. Tellus 45A:311–324

Kurkina O, Talipova T, Pelinovsky E, Soomere T (2011) Mapping the internal wave field in the Baltic Sea in the context of sediment transport in shallow water. J Coastal Res, Special Issue 64:2042–2047

Lepparanta M, Myrberg K (2009) Physical oceanography of the Baltic Sea. Springer, Heidelberg, 378 p

Levitus S (1994) World ocean atlas 1994. Nat Oceanogr Data Center, Inf Rep N 13:47

Morozov EG, Marchenko AV (2012) Short-period internal waves in an arctic Fjord (Spitsbergen). Fizika Atm i Okeana 48(4):461–470 (in Russian)

Paka V, Golenko N, Korzh A (2006) Distinctive features of water exchange across the Słupsk Sill (a full-scale experiment). Okeanologiya 48(S):37–54 (in Russian)

Pelinovsky EN, Polykhin NW, Talipova TG (2000) Geographical and seasonal patterns of the kinematic speed of internal waves in World Ocean. Izv Akad Ing Nauk, Prik Matematika i Informatika 1:133–143 (in Russian)

Piechura J, Walczowski W, Beszczyńska-Möller A (1997) On the structure and dynamics of the water in the Słupsk Furrow. Oceanologia 39:35–54

Piechura J, Beszczyńska-Möller A (2004) Inflow waters in the deep regions of the southern Baltic Sea—transport and transformations. Oceanologia 46(1):113–141

Rak D, Wieczorek P (2012) Variability of temperature and salinity over the last decade in selected regions of the southern Baltic Sea. Oceanologia 54(3):339–354

Reissmann JH, Burchard H, Feistel R, Hagen E, Lass H-U, Mohrholz V, Nausch G, Umlauf L, Wieczorek G (2009) Vertical mixing in the Baltic Sea and consequences for eutrophication—a review. Prog Oceanogr 82(1):47–80

Talipova TG, Pelinovsky EN, Kõuts T (1998) Kinematic characteristics of internal wave field in the Gotland Deep of the Baltic Sea. Okeanologiya 38(1):37–46 (in Russian)

Thorpe SA (2005) The Turbulent Ocean. Cambridge University Press, Cambridge, 484 p

Zhurbas V, Elken J, Paka V, Piechura J, Chubarenko I, Väli G, Golenko N, Shchuka S (2011) On the possibility of convective overturning in the Słupsk Furrow overflow of the Baltic Sea. Oceanologia 53(3):771–791

Appendix A
Sea Water Density According to UNESCO Formula

Density of sea water for a given temperature T in the range $0 < T < 40\,°\mathrm{C}$, salinity S in the range $0 < S < 42\,\mathrm{PSU}$ and pressure p is determined from the equation of state (UNESCO 1981):

$$\rho = \frac{\rho(S,T,0)}{1 - \dfrac{p}{K(S,T,p)}} \tag{A.1}$$

where $K(T,S,p)$ is module of sea water compressibility.

Sea water density is determined using the following algorithm:

- calculation of the SMOW density (*Standard Mean Ocean Water*):

$$\rho_{SMOW} = a_0 + a_1 T + a_2 T^2 + a_3 T^3 + a_4 T^4 + a_5 T^5 \tag{A.2}$$

where:

$$\left.\begin{aligned} a_0 &= 999.842\,594 \\ a_1 &= 6.793\,953 \times 10^{-2} \\ a_2 &= -9.095\,290 \times 10^{-3} \\ a_3 &= 1.001\,685 \times 10^{-4} \\ a_4 &= -1.120\,083 \times 10^{-6} \\ a_5 &= 6.536\,332 \times 10^{-9} \end{aligned}\right\} \tag{A.3}$$

- calculation of sea water density at the normal atmospheric pressure (*normal atmosphere*) ($p = 0$):

$$\rho(S,T,0) = \rho_{SMOW} + B_1 S + C_1 S^{1.5} + d_0\, S^2 \tag{A.4}$$

with:

$$B_1 = b_0 + b_1 T + b_2 T^2 + b_3 T^3 + b_4 T^4 \tag{A.5}$$

S.R. Massel, *Internal Gravity Waves in the Shallow Seas*,
GeoPlanet: Earth and Planetary Sciences, DOI 10.1007/978-3-319-18908-6

$$C_1 = c_0 + c_1 T + c_2 T^2 \tag{A.6}$$

where:

$$\left.\begin{aligned} b_0 &= 8.2449 \times 10^{-1} \\ b_1 &= -4.0899 \times 10^{-3} \\ b_2 &= 7.6438 \times 10^{-5} \\ b_3 &= -8.2467 \times 10^{-7} \\ b_4 &= 5.3875 \times 10^{-9} \end{aligned}\right\} \tag{A.7}$$

$$\left.\begin{aligned} c_0 &= -5.7246 \times 10^{-3} \\ c_1 &= 1.0227 \times 10^{-4} \\ c_2 &= -1.6546 \times 10^{-6} \\ d_0 &= 4.8314 \times 10^{-4} \end{aligned}\right\} \tag{A.8}$$

- determination of the compressibility module at pressure $p = 0$:

$$K(S, T, 0) = K_w + F_1 S + G_1 S^{1.5} \tag{A.9}$$

coefficients K_w, F_1 and G_1 are given by:

$$K_w = e_0 + e_1 T + e_2 T^2 + e_3 T^3 + e_4 T^4 \tag{A.10}$$

where:

$$\left.\begin{aligned} e_0 &= 19\,652.210\,000 \\ e_1 &= 148.420\,600 \\ e_2 &= -2.327\,105 \\ e_3 &= 1.360\,477 \times 10^{-2} \\ e_4 &= -5.155\,288 \times 10^{-5} \end{aligned}\right\} \tag{A.11}$$

$$F_1 = f_0 + f_1 T + f_2 T^2 + f_3 T^3 \tag{A.12}$$

where:

$$\left.\begin{aligned} f_0 &= 54.674\,600 \\ f_1 &= -0.603\,459 \\ f_2 &= 1.099\,870 \times 10^{-2} \\ f_3 &= -6.167\,000 \times 10^{-5} \end{aligned}\right\} \tag{A.13}$$

and

$$G_1 = g_0 + g_1 T + g_2 T^2 \tag{A.14}$$

where:

$$\left.\begin{aligned} g_0 &= 7.9440 \times 10^{-2} \\ g_1 &= 1.6483 \times 10^{-2} \\ g_2 &= -5.3009 \times 10^{-4} \end{aligned}\right\} \quad (A.15)$$

- determination of final compressibility module of the sea water:

$$K(S, T, p) = K(S, T, 0) + A_1 p + B_2 p^2 \quad (A.16)$$

with:

$$A_1 = A_w + (i_0 + i_1 T + i_2 T^2)\, S + j_0\, S^{1.5} \quad (A.17)$$

$$A_w = h_0 + h_1 T + h_2 T^2 + h_3 T^3 \quad (A.18)$$

where:

$$\left.\begin{aligned} h_0 &= 3.239\,90 \\ h_1 &= 1.437\,13 \times 10^{-3} \\ h_2 &= 1.160\,92 \times 10^{-4} \\ h_3 &= -5.779\,05 \times 10^{-7} \end{aligned}\right\} \quad (A.19)$$

$$\left.\begin{aligned} i_0 &= 2.283\,80 \times 10^{-3} \\ i_1 &= -1.098\,10 \times 10^{-5} \\ i_2 &= -1.607\,80 \times 10^{-6} \\ j_0 &= 1.910\,75 \times 10^{-4} \end{aligned}\right\} \quad (A.20)$$

and

$$B_2 = B_w + (m_0 + m_1 T + m_2 T^2)\, S \quad (A.21)$$

where:

$$B_w = k_0 + k_1 T + k_2 T^2 \quad (A.22)$$

with:

$$\left.\begin{aligned} k_0 &= 8.509\,35 \times 10^{-5} \\ k_1 &= -6.122\,93 \times 10^{-6} \\ k_2 &= 5.278\,70 \times 10^{-8} \end{aligned}\right\} \quad (A.23)$$

$$\left.\begin{aligned} m_0 &= -9.9348 \times 10^{-7} \\ m_1 &= 2.0816 \times 10^{-8} \\ m_2 &= 9.1697 \times 10^{-10} \end{aligned}\right\} \quad (A.24)$$

- final determination of the sea water density $\rho(S, T, p)$ from Eq. (A.1):

Example 1: salinity $S = 8$ PSU, temperature $T = 10\,^\circ$C, pressure $p = 0$ bar.
As the pressure $p = 0$, the density $\rho(S, T, 0)$ is given by Eq. (A.4). Thus we have:

- $\rho_{SMOW} = 999.702\,09$ from Eq. (A.2),
- $\rho_w(8, 10, 0) = 1005.946\,59$ from Eq. (A.4).

Example 2: salinity $S = 8$ PSU, temperature $T = 10\,^\circ$C, pressure $p = 10$ bar.
Similarly to example above we obtain:

- $\rho_{SMOW} = 999.702\,09$ from Eq. (A.2),
- $\rho(S, T, 0) = 1005.946\,59$ from Eq. (A.4),
- $K_w = 20\,916.794\,90$ from Eq. (A.10),
- $A_1 = 3.285\,74$ from Eq. (A.17),
- $B_2 = 0.000\,20$ from Eq. (A.21),
- $K = 21\,351.408\,20$ from Eq. (A.16),
- $\rho(8, 10, 10) = 1006.417\,97$ from Eq. (A.1).

Reference

UNESCO (1981) Tenth report of the joint panel on oceanographic tables and standards. UNESCO Technical Papers in Marine Science, Paris, 25 p

Appendix B
Determination of Isopycnal Displacements by Moving CTD Sensors

The method of up and down movement of the towed CTD sensor has been applied for transects in the Southern Baltic waters for many years. When a ship r/v *Oceania* moves with velocity C_s, the CTD sensor falls from the sea surface to the sea bottom and rises from bottom to surface, and again moves down to the sea bottom in the repeated cycles (see Fig. B.1). At each station, when CTD sensor was on the sea surface (or on the sea bottom), the latitude φ and longitude λ, as well as the time t were known. Between two following stations, the ship velocity was only approximately constant. To find the local ship velocity, let us transform the geographic coordinates (φ, λ) into local Cartesian coordinates (x, y), where x is coordinate along east-west direction and y is coordinate along south-north direction.

As the Earth is not a perfect sphere, the radius of the main circle inclined at $\varphi = 55.25°$ becomes:

$$R(\varphi) = \sqrt{\frac{(a^2 \cos\varphi)^2 + (b^2 \sin\varphi)^2}{(a \cos\varphi)^2 + (b \sin\varphi)^2}} \tag{B.1}$$

where $a = 6378.1370$ km is an equator radius and $b = 6356.7523$ km is a polar radius. For $\varphi = 55.25°$ from Eq. (B.1) we obtain $R(55.25°) = 6363.7398$ km. Using the $R(55.25°)$ value we obtain the following relationships between geographical and local Cartesian coordinates:

$$\Delta\lambda = 1° = \frac{\pi \cdot R(55.25°) \cdot \cos(55.25°)}{180} = 63.309\,\text{km}$$

or

$$\Delta\lambda = 1' = \frac{\pi \cdot R(55.25°) \cdot \cos(55.25°)}{60 \cdot 180} = 1.055\,\text{km}$$

S.R. Massel, *Internal Gravity Waves in the Shallow Seas*,
GeoPlanet: Earth and Planetary Sciences, DOI 10.1007/978-3-319-18908-6

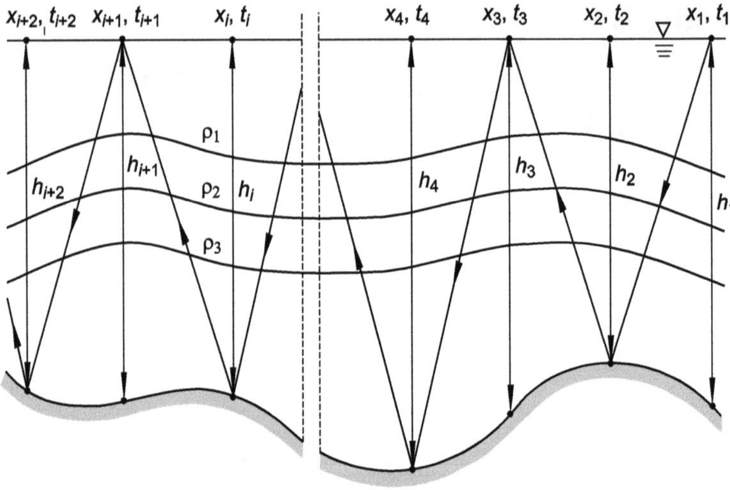

Fig. B.1 CTD sensor zig-zag movement through the water column

and

$$\Delta\varphi = 1^\circ = \frac{\pi b}{180} = 110.946\,\text{km}$$

or

$$\Delta\varphi = 1' = \frac{\pi b}{60 \cdot 180} = 1.849\,\text{km}$$

The ship moved approximately along latitude $\varphi = 55.25^\circ$ and the deviation from this line was around $\pm 3'$ during the whole transect. Therefore for simplicity we neglect this deviation and assume that the transect coincides with the latitude $\varphi = 55.25^\circ$.

The mean local ship velocity between stations i and $i + 1$ is:

$$C_{i,i+1} = \frac{\Delta\lambda}{\Delta t} = \frac{1055 \cdot (\lambda_{i+1} - \lambda_i)}{60 \cdot (t_{i+1} - t_i)} \tag{B.2}$$

where longitude λ_i is expressed in geographical minutes, time t_i is in minutes, and ship velocity is in m s^{-1}.

To assess the position of the given isopycnal displacements in the water column, values of temperature T, salinity S or density ρ at a given point $P_{i,i+1}$, located between stations i and $i + 1$, are required. Let us first determine distance $x_{P_{i,i+1}}$ and time $t_{P_{i,i+1}}$, corresponding to the situation when CTD sensor reaches a selected point $P_{i,i+1}$ with required density ρ, moving down from the sea surface. From geometrical relationships (see Fig. B.2) we have:

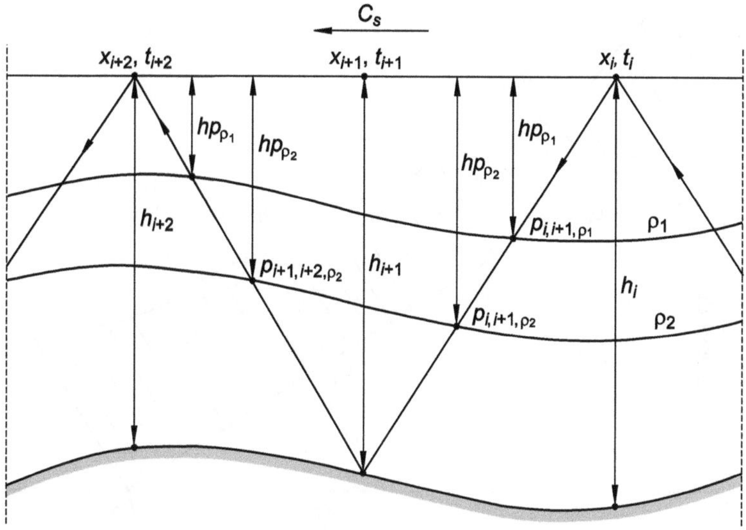

Fig. B.2 CTD sensor crossing the required isopycnal

$$x_{P_{i,i+1}} = \frac{h_{P_{i,i+1}}}{h_{i+1}} (x_{i+1} - x_i) + x_i \tag{B.3}$$

The corresponding time $t_{P_{i,i+1}}$ is given by:

$$t_{P_{i,i+1}} = \frac{h_{P_{i,i+1}}}{h_{i+1}} (t_{i+1} - t_i) + t_i \tag{B.4}$$

Similarly for CTD sensor moving up we obtain:

$$x_{P_{i,i+1}} = x_{i+1} - \frac{h_{P_{i,i+1}}}{h_i} (x_{i+1} - x_i) \tag{B.5}$$

and for time $t_{P_{i,i+1}}$ we have:

$$t_{P_{i,i+1}} = t_{i+1} - \frac{h_{P_{i,i+1}}}{h_i} (t_{i+1} - t_i) \tag{B.6}$$

Using formulae given in the Eqs. (B.3)–(B.6) we can determine functions $T(x, z, t)$ and $S(x, z, t)$, as well as isopycnals $\rho(x, z, t)$ along the experimental transect.

In general, the depth of submergence of a given isopycnal of density ρ is a function of x and t, i.e. $h_\rho = h_\rho(x, t)$. However, when CTD sensor is moving with approximately constant speed C_s, both variables x and t are related as $x = C_s t$. Let us consider the case of a ship moving westward from some point P_0 towards the Słupsk Sill (see Fig. 7.1). The information on the isopycnal displacements can

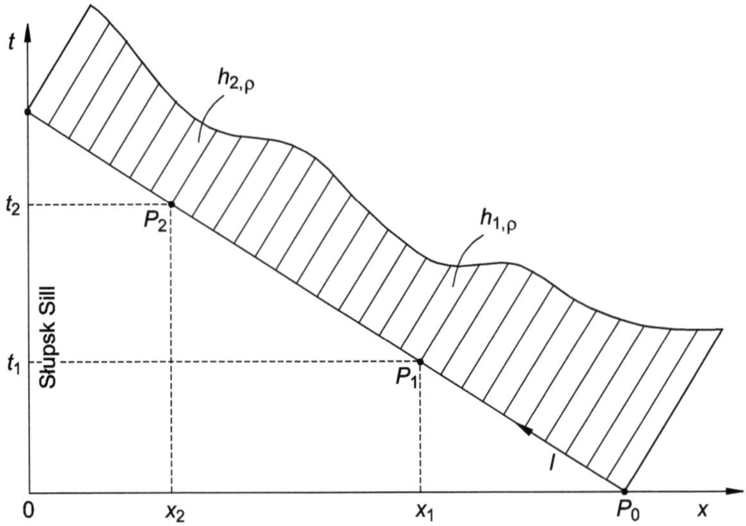

Fig. B.3 Changing of isopycnal displacements along the transect in the two dimensional space (x, t)

be obtained along the cross-section where $x = C_s\, t$. This is schematically illustrated in Fig. B.3. At distance x_1 and corresponding time t_1 the ship is at point P_1 where the given isopycnal ρ is located at water depth $h_{1,\,\rho}$. Similarly, at point P_2 we have isopycnal submergence $h_{2,\,\rho}$. To keep uniform dimensions on both axes, we define the distance l along the cross-section in the two-dimensional space $h = f(x, t)$ as follows:

$$l = \sqrt{[x^2 + (C_s t)^2]} \qquad \text{(B.7)}$$

It should be noted that dependence of the isopycnal submergence h_ρ on time t is validated by the assumption that internal waves profile in the form:

$$\zeta \approx \exp[i(kx \pm \omega t)] = \exp[ik(x \pm C_0\, t)] \qquad \text{(B.8)}$$

where C_0 is the wave phase speed and sign $(\pm)$ reflects the possible different wave directions. Using fact that $x = C_s\, t$, we obtain:

$$\zeta \sim \exp[ik(C_s \pm C_0)t] \qquad \text{(B.9)}$$

For long waves reproduced by solitary waves or by cnoidal waves, the isopycnal displacements are functions of $\cosh(x_1)$ or $cn(x_1)$, where the arguments $x_1 \sim \alpha(x - Ut)$. When the sensor is moving with speed C_s, the arguments become $x_1 \sim \alpha(\pm C_s - U)t$, where parameter α depends on wave amplitude and water depth.

Index

S.R. Massel, *Internal Gravity Waves in the Shallow Seas*,
GeoPlanet: Earth and Planetary Sciences, DOI 10.1007/978-3-319-18908-6

MIX
Papier aus verantwortungsvollen Quellen
Paper from responsible sources
FSC® C105338

If you have any concerns about our products,
you can contact us on
ProductSafety@springernature.com

In case Publisher is established outside the EU,
the EU authorized representative is:
Springer Nature Customer Service Center GmbH
Europaplatz 3, 69115 Heidelberg, Germany

Printed by Libri Plureos GmbH
in Hamburg, Germany